# Heat Exchanger Equipment
# Field Manual

# Heat Exchanger Equipment Field Manual

## Common Operating Problems and Practical Solutions

Maurice Stewart
Oran T. Lewis

AMSTERDAM • BOSTON • HEIDELBERG • LONDON
NEW YORK • OXFORD • PARIS • SAN DIEGO
SAN FRANCISCO • SINGAPORE • SYDNEY • TOKYO

Gulf Professional Publishing is an imprint of Elsevier

Gulf Professional Publishing is an imprint of Elsevier
225 Wyman Street, Waltham, MA 02451, USA
The Boulevard, Langford Lane, Kidlington, Oxford, OX5 1GB, UK

Notices
Knowledge and best practice in this field are constantly changing. As new research and
experience broaden our understanding, changes in research methods, professional practices,
or medical treatment may become necessary.

Practitioners and researchers must always rely on their own experience and knowledge in
evaluating and using any information, methods, compounds, or experiments described
herein. In using such information or methods they should be mindful of their own safety and
the safety of others, including parties for whom they have a professional responsibility.

To the fullest extent of the law, neither the Publisher nor the authors, contributors, or editors,
assume any liability for any injury and/or damage to persons or property as a matter of
products liability, negligence or otherwise, or from any use or operation of any methods,
products, instructions, or ideas contained in the material herein.

Library of Congress Cataloging-in-Publication Data
Heat exchanger equipment field manual : common operating problems and practical
solutions / Maurice Stewart, Oran T. Lewis.
    p. cm.
  Includes index.
  ISBN 978-0-12-397016-9
  1. Heat exchangers–Maintenance and repair.
  TJ263.H3884 2012
  621.402'5–dc23

                                                                                    2011052683
British Library Cataloguing-in-Publication Data
A catalogue record for this book is available from the British Library.

ISBN: 978-0-12-397016-9

For information on all Gulf Professional Publishing
publications visit our Web site at www.elsevierdirect.com

12   13   14   15   16   10   9   8   7   6   5   4   3   2   1
Printed in the United States of America

# Contents

# Heat Transfer Theory

**1**

▶ **OBJECTIVES**

Understand the role of **heat exchangers** (HEX) in upstream
   operations
Be familiar with
   What information is required to
      Define the service and range of operation
      Develop stream properties
   Evaluating heat exchanger operation and performance
   Principles in heat exchanger design
   Basic mechanical standards covering HEX design
      ASME Section VIII, Division I—for most exchangers
      TEMA, Class "R"—for petroleum processing facilities
      API Standards 660, 661, 662, and 632
   What TEMA type is appropriate for each type of service
   How to handle known design/modification challenges
   Know what information is required for heat exchanger
      design and evaluation
   Basic maintenance practices for heat exchangers
This chapter discusses
   Basic heat transfer theory
   Process heat duty
   Heat exchanger nomenclature
   Heat exchanger types and configurations
      Shell and tube
      Double pipe
      Multitube hairpin
      Plate fin (brazed aluminum)
      Plate and frame
      Indirect-fired heaters
      Direct-fired heaters

Waste heat recovery
Air-cooled exchangers
Cooling towers

## ▶ WHAT IS A HEAT EXCHANGER?

Device built for efficient heat transfer from one medium to
   another, whether the media are
   Separated by a wall so they never mix or
   In direct contact
Widely used in both upstream and downstream facilities
May be classified according to their flow arrangement
   Cocurrent flow (parallel flow)
      Two fluids enter the exchanger at the same end and
         travel parallel to one another to the other side
   Countercurrent (counterflow)
      Two fluids enter the exchanger from opposite ends

### Commonly Used Types of Heat Exchangers

Shell and tube
   Consist of a bundle of tubes and a shell
   One fluid that needs to be heated or cooled flows through
      the tubes and the second fluid runs over the tubes that
      provides the heat or absorbs the heat required
Plate
   Composed of multiple, thin, slightly separated plates that
      have very large surface areas relative to size for heat
      transfer
Plate fin
   Uses fins to increase the efficiency of the unit
   Designs include cross-flow and counterflow coupled
      with various fin configurations (straight, offset, or
      wavy)

Air cooled
  A bundle of tubes with air passing around the outside

## Uses of Heat Exchangers
Heat and cool fluids
  Heat recovery
    When not economical, an air cooler is used frequently
  Separation
    Reboiling
    Condensing

## Heat Exchangers—The Bad News
Leading cause of **lost production opportunities (LPO's)**—
  see Figure 1.1

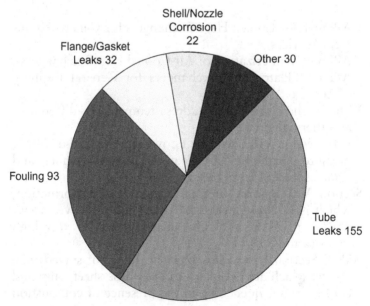

**Figure 1.1** Typical causes of heat exchanger LPO's (based on 332 cases in a 5-year study; courtesy of Chevron USA).

Are very susceptible to problems, despite having no moving parts

Problems include fouling, corrosion, and leaks

Can limit unit throughput

## General Considerations

Starting point for heat exchanger design

Heat exchangers are designed to ASME Boiler and Pressure Vessel Codes

Section VIII Division 1 for most exchangers

Section I for "fired" steam generators

Standards of the Tubular Exchanger Manufacturers Association (TEMA)—Class "R" for petroleum processing

API standards:

API 660 Shell and Tube Exchangers for General Refinery Services

API 661 Air-Cooled Heat Exchangers for General Refinery Services

API 632 Winterization of Air-Cooled Heat Exchangers

API 662 Plate Heat Exchangers for General Refinery Services

None of these standards address temperature differences and thermal stresses

ASME Section VIII, Division 1 covers mechanical design of pressure containing parts (e.g., shells and channel) and some aspects of large flanges

Section VIII ignores thermal stresses and deformation, which is of consequence to flange tightness over about 250°F or 121°C, and tube sheet integrity at very high temperatures

ASME Section I provides rules for design of stayed tube sheets, which are necessary to keep tube sheets thin and cool by water quenching in the presence of combustion or other hot gases

TEMA address shell and tube exchangers only and cover nomenclature, fabrication tolerances, standard clearances, minimum plate thickness, and tube sheet design rules

API Standards reference the aforementioned standards and several others relating to piping, welding materials, and nondestructive examination; requires many design features that are universally applicable (e.g., seal bars) and a checklist of decisions that must be made by the purchaser

Most companies use API 660 for shell and tube exchangers and API 661 for air-cooled exchangers

There is little demand and incentive to use plate exchangers in downstream refinery service

Industry standards are limited to moderate pressures and temperatures and do not cover subjects that are beyond the control of manufacturers, such as fouling, corrosion, vibration, leak tightness of flanges, and tube ruptures

## Company Engineering Standards and Specifications

Address subjects that are applicable to most exchangers and that are **beyond the scope** of industry standards

References and supplements API 660 for shell and tube exchangers and addresses additional requirements concerning

Flange design

Tube vibration

Tube rupture

Covers exchangers using special materials or thick wall construction

References and supplements API 661 for air coolers

Covers double pipe and hairpin exchangers

Covers plate and frame exchangers

Ensures optimum design and eliminates guesswork

Provides additional information for special cases

High-pressure closures
Waste heat boilers
Acoustic vibration

## Flanges

Industry design standards call for flanges that are stronger than ASME requirements.

Flange design has evolved through several phases during the past 50 years
  Industry in 1960s
    Successfully passed hydrotest
    50% leak in service (high temperature, low pressure)
    Not economical to correct (steam rings)
  Industry in 1970s
    Recognized the need for fundamental change
    Testing, 30+ sponsors
  Companies in 1980s until today
    Use beefy flanges
    Environmentally driven

## Tube Vibration

Take steps during design to prevent tube vibration
As with flange design, the attention devoted during design to tube vibration has evolved over time
A key part of preventing vibration is determining which tubes will vibrate
Other design steps include
  Bundle layout
  Shell-side velocity
  Unsupported length
Prevention of tube vibration must be integrated with layout, process design, and mechanical design concerns

See Figures 1.2 and 1.3.

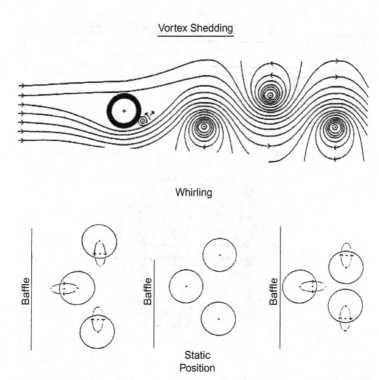

Figure 1.2 Tube vibration mechanisms.

## Tube Rupture

Most companies' design standards improve on the "2/3 rule"

Design standards have evolved

   Investigated catastrophic plant incidents

   Large scale laboratory tests by Southwest Research Institute 1977–1979

   Computer modeling (method of characteristics)

   Developed simple accurate analytical methods

     Rupture flow and maximum pressure controlled by compression wave speed in low-pressure piping

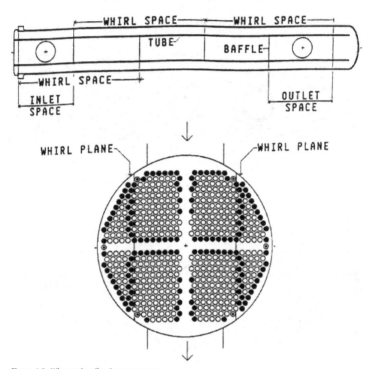

**Figure 1.3** Where tube vibrations may occur.

Tube rupture industry standards

ASME Section VIII Division 1 UG-133 (d) Code case VIII-80-56 (6/25/80) requires relief provisions; workable in steam boilers only

API recommended practice 521 discusses inadequacy of ASME code rule and is not helpful

Ten-thirteenths (10/13) rule

LP side MAWP = 10/13 × HP side MAWP

### Tube Rupture Design Considerations

Consider complete break of one tube

Consider break at one end (worst case)

Treat high-pressure two-phase fluid as all gas (liquid is moving slower)

Treat HP shell and LP channel same as HP channel and LP shell

See Figure 1.4.

### High-Pressure Closures

High-pressure closures are not commodity items

High-pressure closures must be custom designed

These closures are not easy to design

Use a manufacturer with a track record

Use a high-pressure closure when

Pressure (psi) × diameter (in.) is greater than 75,000

Pressure (bar) × diameter (mm) is greater than 131,345

## Acoustic Vibration

Rare but must be eliminated when it does occur

Can occur but may not damage tubes

Do not use a 45° layout for gas service

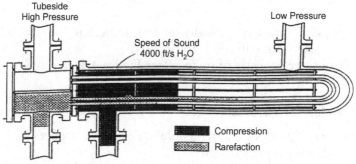

**Figure 1.4** Tube rupture transient schematic.

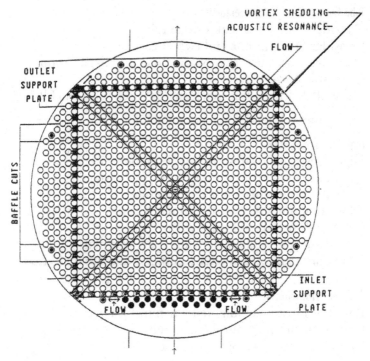

**Figure 1.5** Acoustic resonance.

Acoustic vibration is usually ignored in design
   A noise—but still needs to be fixed
   Consider acoustic resonance (see Figure 1.5)


▶ **PROCESS SPECIFICATION**

Describes process requirements and design preferences for a
   heat exchanger's design
Communication tool between the company and the heat exchanger manufacturer

Usually includes a filled-out
  API 660 specification sheet or company design data sheet
  API 660 checklist
  Additional items not addressed on either form
Used to
  Define a new heat exchanger service
  Provide sufficient information to analyze an existing
    exchanger
  Information flows in both directions
  Provides information the manufacturer needs from the
    company
  Company receives the following from the manufacturer
    Mechanical design information, including fabrication
      drawings
    Heat transfer and pressure drop performance infor-
      mation
    Vibration analysis results

## ▶ PROCESS SPECIFICATION SHEET VS. DESIGN DATA SHEET

### Process Specification Sheet
Name of a standard sheet (included in API 660 Appendix
  D) used for HEX specifications
Describes processes and preferences
Key fields where service is defined are
  Performance of one unit section
  Remarks section
Performs similar function to design data sheet

### Design Data Sheet
Describes processes and preferences
Lists key parameters needed for
  Design of a new service

Analysis of an existing exchanger
Key data fields to define service are
Performance data section
Notes section

## ▶ INFORMATION NEEDED FOR SPECIFYING WORK

### Heat Exchanger Design
General information
Performance requirements
Stream data
Materials of construction
Mechanical considerations
Preferences
Deliverables

### Confirm with the Supplier That
Your package is complete
What deliverables are expected and the delivery date
How much services will cost

## ▶ DELIVERABLES FROM SUPPLIER

### Formal Deliverables Ensure Full Completion of the Project
API spec sheet
HTRI simulations
Drawings
Report
Price/fabrication schedule
Invoice for services

## ▶ EVALUATING DESIGNS

### Based on Process Specification Sheets

Received process specification sheet from supplier

Verify information—this is the company's feedback loop

Make certain the form is filled out properly, or are there many of the numbers "back-calculated"

For each design one must ensure they understand

Basis for fundamentals

Reasons behind the service specifications

**CAUTION:** One cannot just design to industry standards

### Required Physical/Thermal Properties

Dependent on the phase(s) for each stream

Use standard form from API 660 Appendix D and then add additional information

Areas of concern include:

Performance requirements

Stream data

For each phase or for single-phase fluids the following is required:

Density lb/ft$^3$ (kg/m$^3$)

Viscosity cp (Pa s)

Specific heat (capacity) Btu/(lb °F)[J/(kg °C)]

Thermal conductivity Btu/(hr °F ft)[W/(m °C)]

Over the range extending between inlet to inlet temperature of two fluids through exchanger or range where the phase exists

For boiling fluid the following is required:

Vapor molecular weight

Weight fraction vapor

Liquid and vapor phase enthalpies (RTF-2)

Boiling range (bubble to dew point)

Critical temperature and pressure

Inlet thermal conductivity based on inlet temperature
Btu/(hr °F ft) (W/(m °C))

Surface tension over the range extending between inlet
temperatures to dew point

Usually at two pressures spanning the operation

For condensing fluid, the following is required:

Vapor molecular weight

Weight fraction vapor

Stream enthalpy

Surface tension (especially for finned tubes) over the range
extending between the inlet-to-inlet temperature of the
two fluids through the exchanger *or* through the range
where the phase exists

## Thermal and Hydraulic Design

Heat transfer analysis use basic formulas and computer
programs

Much of the film coefficient information is proprietary

Pressure drop

Single-phase tube-side straightforward

Every other mode, shell side or phase change, is difficult

Generally obtained from proprietary sources

An important relationship is one between heat transfer and
pressure drop

Also need to consider system economics

Allocate the streams to minimize equipment cost and operating cost

Mechanical and materials constraints include:

Minimizing

Initial equipment cost [capital expense (OPEX)] and
Operating cost [operating expense (OPEX)]

Thermal expansion

Flange and tube leaks

Safety
Materials versus process conditions
Fouling information is highly proprietary
    Need to understand and control fouling so as to avoid its
      high costs

## ▶ PRESSURE DROP CONSIDERATIONS

### Key Parameter for Design and Troubleshooting

Pressure drop should be set by economics
Key pressure drop considerations
    Evaluating design
      Determine flow path length
      Calculate pressure gradient (DP/L)
      What is reasonable pressure drop? Depends on service.
        Minimize cost system. Restricted? Why? Impact lim-
        iting side
      Impact fouling
    Evaluating performance:
      Compare to design
      Another indicator of fouling
      What is time trend?
How to adjust pressure drop
    Tube side
      Number of tubes per pass
      Tube length
      Number of tube passes
      Number of shells and routing
    Shell side
      Flow path length (TEMA type, tube length)
      Baffles (style, cut, spacing)
      Number of shells and routing

Other ways
  Install tube inner diameter inserts
  Change tube pitch
For single-phase and two-phase (still predominantly trans-
  ferring sensible heat) streams, the drop is simply calculated by multiplying the economic pressure gradient by
  the appropriate flow path length as shown in the following equation:

$$\Delta P \ (psi) = \frac{\Delta P}{L}(L) \qquad (1.1)$$

The resulting pressure drop is for the overall exchanger, in-
  cluding nozzles, channels, heads, or U bends
For two-phase streams, the density is the homogeneous den-
  sity assuming no liquid holdup
Pressure gradients ($\Delta P/L$)
  For liquids (HC and $H_2O$):
    Tube side: 0.2 to 0.3 psi/ft (4524 to 6784 Pa/m)
    Shell side: 0.4 to 0.6 psi/ft (9048 to 13570 Pa/m)
  For all fluids including two phase and gases:
    Tube side: (0.05 to 0.08) × (density, $lb/ft^3$)$^{1/3}$ (1131 to
      1809) * [(density, $kg/m^3$)$^{1/3}$]/2.52 (Pa/m)
    Shell side: (0.1 to 0.15) × (density, $lb/ft^3$)$^{1/3}$ (2262 to
      1809) * [(density, $kg/m^3$)$^{1/3}$]/2.52 (Pa/m)
Flow path lengths, ft or m (L)
  Tube side: tube straight length × number of tube passes
  Shell side: axial length (for E shell tube straight length)
Pressure drop can be estimated for new process conditions
  based on the Darcy–Weisbach flow equation. The equa-
  tion, expressed in feet of liquid, is expressed as follows:

$$h = \frac{fL}{D}\frac{V^2}{2g} = \left(\frac{fL}{D} + K\right)\left(\frac{V^2}{2g}\right) \qquad (1.2)$$

where

$h$ = head loss, ft (m)
$f$ = Darcy friction factor
$L$ = pipe length, ft (m)
$D$ = pipe inside diameter, ft (m)
$V$ = fluid velocity, ft/s (m/s)
$g$ = gravitational constant (32.17 ft/s$^2$ or 9.807 m/s$^2$)
$K$ = fitting loss coefficient

The Darcy–Weisbach equation expressed in psi (Pa) is as follows:

$$P = \frac{fL}{D} \; \frac{m^2}{\rho A^2 (1.2 \times 10^{11})}$$
$$= \frac{fL}{D} \frac{m^2}{\rho D^4 (7.4 \times 10^{10})} \tag{1.3}$$

where

$P$ = pressure drop, psi (Pa)
$m$ = mass flow rate, lbm/hr (kg/s)
$\rho$ = fluid density, lbm/ft$^3$ (kg/m$^3$)
$A$ = flow area cross section, ft$^2$ (m$^2$)

For SI units, the Darcy–Weisbach equation is as follows:

$$P = \frac{fL}{D} \frac{m^2}{\rho A^2} \tag{1.4}$$

For both sides, pressure drop is proportional to density and to the flow rate (or velocity) squared

This requires a known pressure drop with corresponding operating conditions to predict operation at new conditions

Note that this normalizing technique works only when
   there is no change in the flow regime or in shell-side flow
   partitioning (more or less leakage or bypassing)
Because shell-side flow is complex, do not overly
   extrapolate
The Darcy friction factor can be determined from the
   graph in Figure 1.6

Shell-Side Flow Stream
See Figure 1.7.

## ▶ BASIC HEAT TRANSFER THEORY

Heat is that energy transferred solely as a result of a temperature change.

### Heat Transfer Mechanisms
Conduction
The transfer of heat from **one molecule to an adjacent molecule** while the particles remain in fixed positions relative
   to each other
Primary mechanism in **solids** and some **fluids** that are **stag-
   nant** or have low flow rates

Convection
Heat is transferred by the physical movement of molecules
   from place to place, e.g., the mixing of warmer and cooler
   portions of a fluid in a heater
Primary mechanism in fluid–fluid exchangers

Radiation
Process where **heat waves** are emitted that may be absorbed,
   reflected, or transmitted through a colder body
Sun heats the earth by electromagnetic waves
Hot bodies emit heat waves

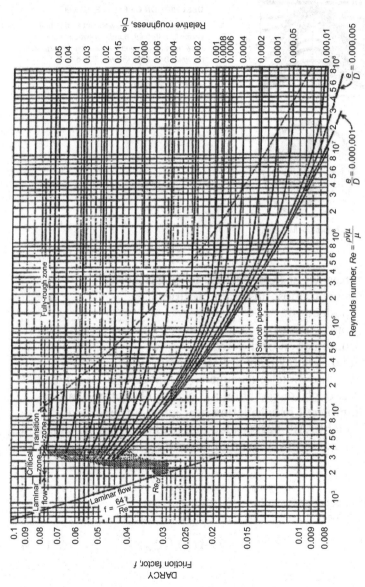

**Figure 1.6** Friction factor for fully developed flow in circular pipes.

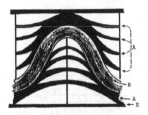

**DESCRIPTION OF FLOW STREAMS**

A.  Flow through annuli between tubes and baffles
B.  Cross-flow through bundle between tips
C.  Flow through annulus between bundle and shell between baffle tips (shown dotted)
D.  Flow through tube pass partition lanes parallel to cross-flow direction

**Figure 1.7** Shell-side flow stream.

## Upstream Oil- and Gas-Handling Facility Applications

In general, most upstream facilities use
  Conduction
  Convection or a
  Combination of the two
Radiant energy from a direct flame
  In fluid–fluid exchangers, temperatures are not hot
    enough for radiation to be a significant mechanism
  Important in calculating heat given off by a flare
  Refer to API 521 "Guidelines for Flare System Sizing and
    Radiation Calculation"

## Basic Equations
Conduction

$$q = k \, (A/L) \, (\Delta t_M) \qquad (1.5)$$

Convection

$$q = hA \, (\Delta t_M) \qquad (1.6)$$

Radiation

$$q = \sigma \, A T^4 \qquad (1.7)$$

where

| | | |
|---|---|---|
| $q$ | = | heat transfer rate, Btu/hr |
| $A$ | = | heat transfer area, ft$^2$ |
| $\Delta t_M$ | = | temperature difference, °F |
| $k$ | = | thermal conductivity, Btu/hr ft °F |
| $h$ | = | film coefficient, Btu/hr ft$^2$ °F |
| $L$ | = | distance heat energy is conducted, ft |
| $\sigma$ | = | Stefan–Boltzmann constant |
| | = | $0.173 \times 10^{-3}$ Btu/hr ft |

The value of "h" is a proportionality constant used to characterize liquid film resistance

The value "k" is the thermal conductivity of the solid separating the two fluids

**Flow of Heat**

If **heat** is being **transferred through various layers** and via **various modes**, two conclusions can be stated regarding flow of heat:

The **heat flow** ($q$) is **equal** through all layers, thus

$$q_1 = q_2 = q_3 \qquad (1.8)$$

The heat flow is equal to the overall temperature difference divided by the total thermal resistance, thus

$$q = \frac{\Delta t_M}{\sum R} \qquad (1.9)$$

where

$R$ = thermal resistance of each layer, hr/°F/Btu

$\quad = \dfrac{L}{kA}$ (conduction)

$\quad = \dfrac{1}{hA}$ (convection)

The $\Sigma R$ is defined as $\left(\dfrac{1}{UA}\right)$ and Equation (1.9) can be rewritten as

$$q = UA\, \Delta t_M \qquad (1.10)$$

where
$U$ = overall heat transfer coefficient, Btu/hr ft$^2$ °F
Equation (1.10) is the basic equation used for heat transfer calculations
Equation (1.10) can be rearranged to solve for the area required, which is the major design parameter for all heat exchanger design calculations

$$A = \frac{q}{U\Delta t_m} \qquad (1.11)$$

In order to calculate the area, **three parameters** must be determined
Mean-temperature difference, $\Delta t_M$
Heat transfer coefficient, $U$
Heat duty, $q$

## Multiple Transfer Mechanisms

Heat exchangers transfer energy in three steps

Convective steps

Hot fluids to exchanger tube

Exchanger tube to cold fluid

Conductive step

Flow of heat through exchanger tube wall

## ▶ DETERMINATION OF MEAN TEMPERATURE DIFFERENCE

## Mean Temperature Difference (MTD)

The driving force for heat transfer is the temperature differences between the two fluid streams

Because the temperature of the process fluid changes as it flows, a "mean" temperature difference must be used

Mean temperature difference

$$MTD = F\,(LMTD) \qquad (1.12)$$

where

$$\mathrm{LMTD} = [(\Delta t_1 - \Delta t_2)/\ln(\Delta t_1/\Delta t_2)] \qquad (1.13)$$

where

$\Delta t_1$ =  largest terminal $\Delta t$

$\Delta t_2$ =  smallest terminal $\Delta t$

ln    =  logarithm to the base e

$F$    =  correction factor for heat exchanger geometry

=  1.0 for pipe-in-pipe and countercurrent

=  1.0 for boiling or condensing of pure components

For exchangers with straight-line heat release curves, the MTD is equal to the calculated log mean temperature difference (LMTD) with the "F" correction factor equal to 1.

Single and two-phase exchangers and some condensers typically fall into this category

LMTD is shown in Figure 1.8.

The expression for the F-correction factor is used for a single TEMA E-shell with an even number of tube passes

$$F = \frac{\eta}{\delta \ln\{[2 - P(1 + R - \eta)]/[2 - P(1 + R + \eta)]\}} \quad (1.14)$$

and

$$\eta = \sqrt{R^2 + 1} \quad (1.15)$$

$$R = \frac{T_i - T_o}{t_o - t_i} \quad (1.16)$$

$$P = \frac{t_o - t_i}{T_i - t_i} \quad (1.17)$$

$$\delta = \frac{R - 1}{\ln[(1 - P)/(1 - PR)]} \quad (R \neq 1) \quad (1.18)$$

$$\delta = \frac{1 - P}{P} \quad (R = 1) \quad (1.19)$$

where
$P$  =  thermal effectiveness
$R$  =  ratio of cold to hot stream flowing heat capacities—flow rate times heat capacity or the slope of the enthalpy curve

When $R$ is 1, different expressions for the LMTD and $F$ correction are used

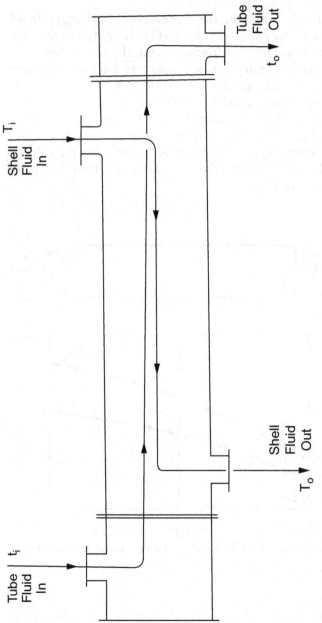

**Figure 1.8** Log mean temperature difference for an exchanger with straight-line heat release curves.

For the *F* correction, the expression for δ changes. In addition, the expression for LMTD simplifies to the average (constant) approach between the two streams

For practical purposes, the minimum F-correction factor to use in design should be about 0.8

Two fluids may transfer heat in either

Countercurrent direction (Figure 1.9) **or**

Concurrent direction (Figure 1.10)

Relative direction of the two fluids influences the value of the LMTD, and thus the area required to transfer a given amount of heat

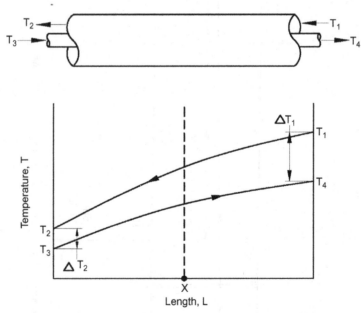

**Figure 1.9** Countercurrent flow ($T_1$, hot fluid in; $T_2$, hot fluid out; $T_3$, cold fluid in; $T_4$, cold fluid out).

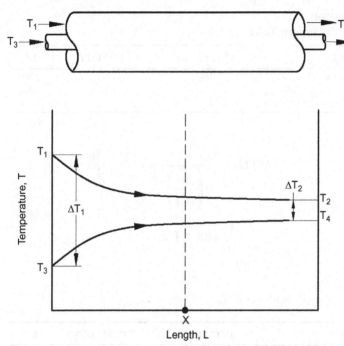

**Figure 1.10** Cocurrent flow ($T_1$, hot fluid in; $T_2$, hot fluid out; $T_3$, cold fluid in; $T_4$, cold fluid out).

## Example 1: Log Mean Temperature Difference Determination

Given:

A hot fluid enters a concentric pipe at a temperature of 300°F and is to be cooled to 200°F by a cold fluid entering at 100°F and heated to 150°F

Determine:

LMTD for

(a) cocurrent flow

(b) countercurrent flow

Solution:

(a) cocurrent flow

| SIDE | HOT FLUID | COLD FLUID | $\Delta T$ |
|---|---|---|---|
| Hot fluid inlet | 300 | 100 | 200 |
| Hot fluid outlet | 200 | 150 | 50 |
| $\Delta t_1 = \text{GTTD} = 200$; $\Delta t_2 = \text{LTTD} = 50$ | | | |

Thus,

$$LMTD = \frac{\Delta t_1 - \Delta t_2}{ln\left(\frac{\Delta t_1}{\Delta t_2}\right)} = \frac{200 - 50}{ln\dfrac{200}{50}}$$

$$= 108.2°F$$

(b) countercurrent flow

| SIDE | HOT FLUID | COLD FLUID | $\Delta T$ |
|---|---|---|---|
| Hot fluid inlet | 300 | 150 | 150 |
| Hot fluid outlet | 200 | 100 | 100 |
| $\Delta t_1 = \text{GTTD} = 150$; $\Delta t_2 = \text{LTTD} = 100$ | | | |

Thus,

$$LMTD = \frac{\Delta t_1 - \Delta t_2}{\ln\left(\frac{\Delta t_1}{\Delta t_2}\right)} = \frac{150 - 100}{\ln\dfrac{150}{100}}$$

$$= 123.3°F$$

Figure 1.11 is a graphical solution for determining LMTD

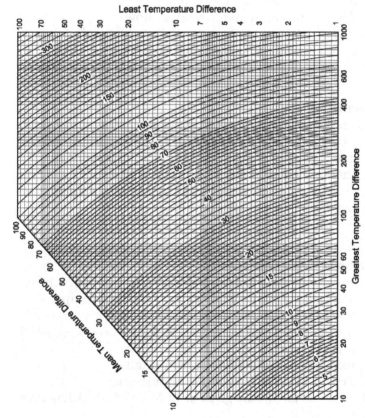

Figure 1.11 Graph for solving LMTD.

## Log Mean Temperature Difference

The assumptions used in the derivation of LMTD are summarized here

Constant "U" throughout exchanger

Smooth-line temperature profiles

True cocurrent or countercurrent flow

These assumptions are not generally valid for the majority of heat exchangers, and corrections must be made

## Nonconstant "U"

Overall heat transfer coefficient **will not remain constant**

Calculations based on a value of U taken **midway between ends** of the exchanger are **usually accurate enough**

If there is a **considerable variation in U** from one end of the exchanger to the other, then a **step-by-step numerical integration** is necessary

## Different Flow Arrangements

More complicated flow arrangements than simple cocurrent and countercurrent exist in heat exchanger equipment

These configurations are more difficult to treat analytically, thus a temperature correction factor is introduced, which is a function of

Exchanger inlet and outlet temperature

Type flow arrangements for both streams

Correction factors are given in **TEMA and GPSA** Engineering Data Book

As a practical matter, an **exchanger** with a correction factor **less than 0.8** should **not be used**

**Additional data** on correction factors are **presented** in the **shell-and-tube** and **air cooler subsections** of this section

## Nonlinear Temperature Profile

Nonlinear temperature profiles occur in three situations (Figure 1.12)

Condensation plus subcooling

Vaporization plus subcooling

Curved T plot vs. Q plot

For purposes of analysis, this **process** may be considered the **super position of two or more exchangers** as was depicted in **Figure 1.10**.

The total heat exchanger area **cannot be calculated** using $A = q/(u) \ \Delta t_M$ when temperature profiles are nonlinear

Summaries of several procedures follow:

1. Calculate the area for each segment of the exchanger. The heat transfer coefficient (U) varies through the exchanger. The correct U for each segment depends on the fluid phase in that segment. The exchanger area is the sum of the segments' areas.

2. If an overall heat transfer coefficient is known, the area can be calculated using a weighted LMTD (WLMTD), expressed as follows:

$$WLMTD = \frac{\sum q}{\dfrac{q_1}{LMTD_1} + \dfrac{q_2}{LMTD_2} + \dfrac{q_3}{LMTD_3}} \qquad (1.20)$$

## Example 2: Application of WLMTD

Given: Figure 1.13

Determine:

Cooling and condensation weighted temperature difference

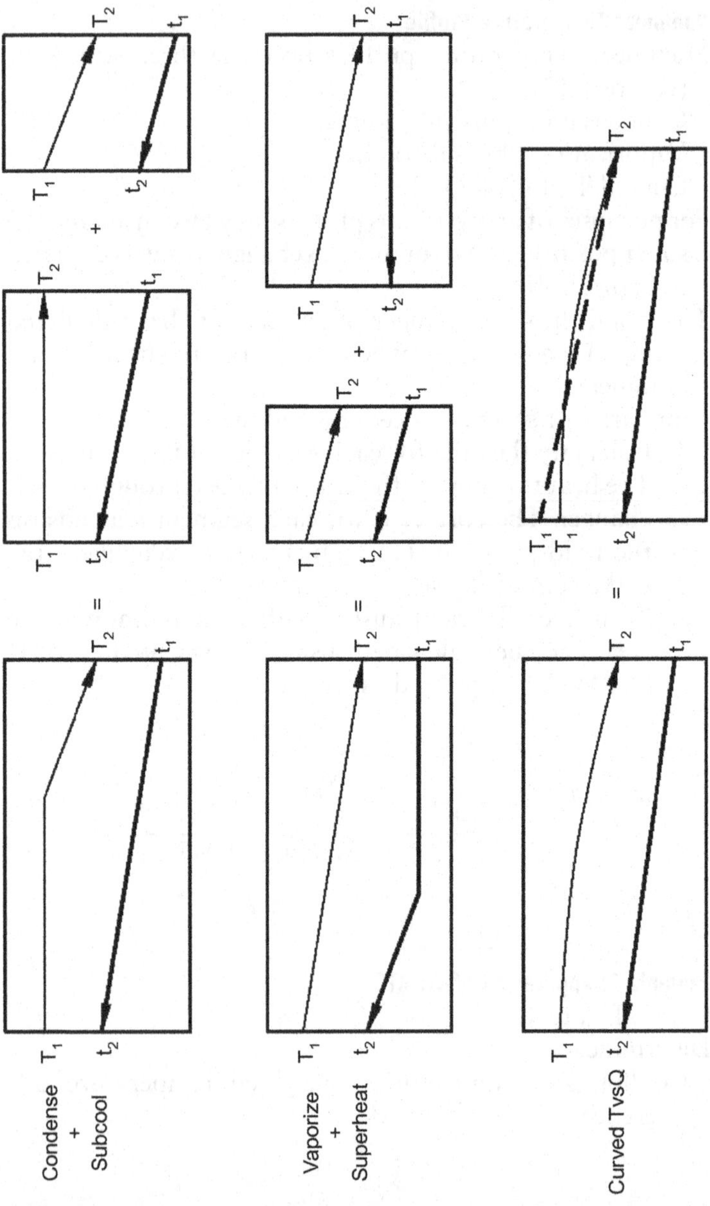

**Figure 1.12** Nonlinear temperature profiles.

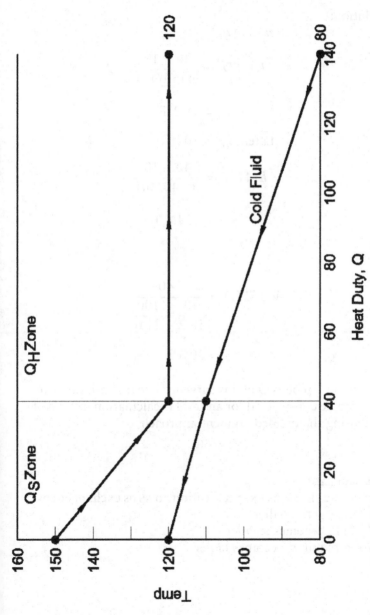

Figure 1.13  LMTD for cooling and condensation.

**Solution:**

$$\textit{Sensible } q_s = 40$$

$$LMTD_s = \frac{30 - 10}{\ln (30/10)}$$

$$= 18.2°F$$

$$\text{Latent } q_h = 100$$

$$LMTD_h = \frac{40 - 10}{\ln (40/10)}$$

$$= 121.6°F$$

Thus,

$$WLMTD = \frac{140}{\dfrac{40}{18.2} + \dfrac{100}{21.6}}$$

$$= 20.5°F$$

3. If the T plot vs Q plot is curved, a linear approximation can be constructed for an LMTD calculation instead of using the divided segment approach

**Approach ($\Delta t_2$)**
Economic choice as its specification governs **exchanger cost**
As "$\Delta t_2$" gets smaller
   LMTD becomes smaller
   Area required becomes larger

Because the cost of an exchanger is a direct function of area, specification of approach has a direct effect on cost

When specifying heat exchangers, it is usually beneficial to specify a maximum or minimum approach to the vendor. This establishes an upper or lower limit, below or above that the actual approach must occur

Common approaches

Aerial coolers (18–45°F)

Water cooling of hydrocarbon liquids and gases (14–22°F)

Liquid–liquid heat exchange (20–45°F)

Refrigeration chillers on gas–liquid streams (7–11°F)

▶ **DETERMINATION OF HEAT TRANSFER COEFFICIENT**

**Overview**

Heat transfer coefficient for a heat exchanger can be defined as the sum of thermal resistances per unit area

$$U = \frac{1}{\sum R_i A} \qquad (1.21)$$

where

U = Overall heat transfer coefficient, Btu/hr ft$^2$ °F

**Area Basis**

Almost all heat exchanger problems involve heat transfer, through three or more layers around circular tubes

Normally there are two areas to consider

Inside tube area ($A_i$)

Outside tube area ($A_o$)

Unless specifically stated otherwise, **all heat transfer values presented** in the literature are **based** on **outside tube area** ($A_o$)

## Heat Transfer Coefficient—Clean Tube

Heat transfers through a clean tube are shown in Figures 1.14 and 1.15.

Equation (1.10), the general heat transfer equation, can be rewritten using the thermal resistances in the three layers

$$R_1 = \text{inside convection resistance} = \frac{1}{h_i A_i}$$

$$R_2 = \text{conduction through tube wall} = \frac{L}{k A_w}$$

$$R_3 = \text{outside convection resistance} = \frac{1}{h_0 A_0}$$

Thus,

$$q = \frac{\Delta t_m}{\sum R}$$

$$= \frac{\Delta t_m}{R_1 + R_2 + R_3} \qquad (1.22)$$

$$= \frac{\Delta t_m}{\dfrac{1}{h_i A_i} + \dfrac{L}{k A_w} + \dfrac{1}{h_0 A_0}}$$

Assuming $A_w$ (**average tube area**)$=A_0$ (**outside tube wall area**), as $L/(k\,A_w)$ for metal tubes is a small number

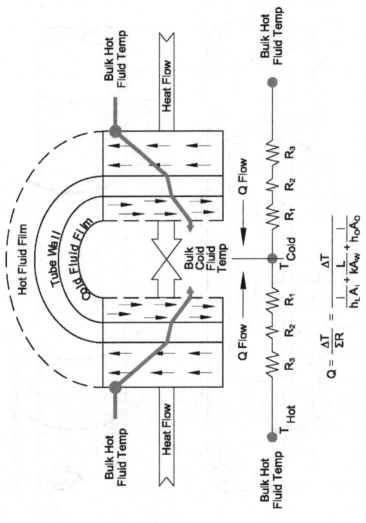

Figure 1.14 Heat transfer through a clean tube.

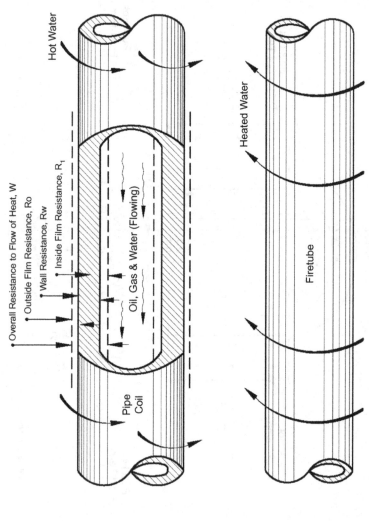

Hot Water

Heated Water

Overall Resistance to Flow of Heat, W
Outside Film Resistance, Ro
Wall Resistance, Rw
Inside Film Resistance, R₁

Oil, Gas & Water (Flowing)

Pipe Coil

Firetube

**Figure 1.15** Heat transfer path involving resistances in series.

compared to the film resistances, Equation (1.15) can be arranged as follows:

$$q = \left[ \frac{1}{\frac{1}{h}\left(\frac{A_0}{A_i}\right) + \frac{L}{k} + \frac{1}{h_0}} \right] A_0 \Delta t_m$$

which is identical in form with Equation (1.10), where

$$U = \frac{1}{\frac{1}{h_i}\left(\frac{A_0}{A_i}\right) + \frac{L}{k} + \frac{1}{h_0}}$$

where
$h_i$ = inside film coefficient, Btu/hr ft$^2$ °F
$h_0$ = outside film coefficient, Btu/hr ft$^2$ °F

## Heat Transfer Coefficient—Fouled Tube
Overview
   **Thin layers** of dirt, scale, corrosion products, or degradation products can **build up over time** on the **inside or outside** of the tube and **reduce** the **rate of heat transfer** (Figure 1.16)
Fouling factors
   Included in calculations to account for film buildup
   ($R_0$ = outside resistance, $R_i$ = inside resistance)
   Presented in the literature for various materials and conditions
Because scale or dirt resistance increases with time in service, some time basis must be chosen for numerical values of fouling factors. The time basis of 1 year is commonly used

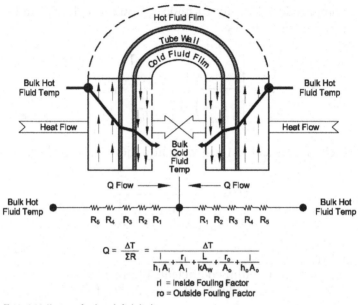

**Figure 1.16** Heat transfer through fouled tubes.

Five thermal resistances must be considered when calculating the heat transfer coefficient for fouled tubes:

$R_1$ = inside convection resistance $= \dfrac{1}{h_i A_i}$

$R_2$ = inside fouling resistance $= \dfrac{r_i}{A_i}$

$R_3$ = conduction through wall $= \dfrac{L}{k A_w}$

$R_4$ = outside fouling resistance $= \dfrac{r_0}{A_0}$

$R_5$ = outside convection resistance $= \dfrac{1}{h_0 A_0}$

Thus

$$q = \frac{\Delta t_m}{\sum R}$$

$$= \frac{\Delta t_m}{\dfrac{1}{h_i A_i} + \dfrac{r_i}{A_i} + \dfrac{L}{k A_w} + \dfrac{r_0}{A_0} + \dfrac{1}{h_0 A_0}} \tag{1.23}$$

Assuming $A_w = A_0$, then

$$q = \left| \frac{1}{\dfrac{1}{h_i}\left(\dfrac{A_0}{A_i}\right) + r_i\left(\dfrac{A_0}{A_i}\right) + \dfrac{L}{k} + r_0 + \dfrac{1}{h_0}} \right| A_0 \Delta t_m \tag{1.24}$$

where

$$U = \frac{1}{\dfrac{1}{h_i}\left(\dfrac{A_0}{A_i}\right) + r_i\left(\dfrac{A_0}{A_i}\right) + \dfrac{L}{k} + r_0 + \dfrac{1}{h_0}} \tag{1.25}$$

where

| | | |
|---|---|---|
| $h_i, h_0$ | = | inside, outside film coefficient, Btu/hr ft$^2$°F |
| $k$ | = | thermal conductivity, Btu/hr ft$^2$°F |
| $A_i, A_0$ | = | inside, outside surface area, ft$^2$/ft |
| $R_i, R_0$ | = | inside, outside fouling resistance, hr ft$^2$°F/Btu |
| | = | 0.003 hr ft$^2$°F/Btu |
| $L$ | = | wall thickness, ft |

**A heat transfer coefficient** based on **fouled tubes** can be determined by knowing the clean heat transfer coefficient and the fouling factor

$$U_{fouled} = \cfrac{1}{\cfrac{1}{U_{clean}} + r_i \left( \cfrac{A_0}{A_i} \right) + r_0} \qquad (1.26)$$

Actual fouling in heat exchangers depends on several characteristics
  Nature of fluid and material deposited
  Fluid velocity
  Temperature of fluid
  Temperature of tube wall
  Tube wall material
  Tube wall surfaces (finned or smooth)
  Time since last cleaning
**Table 1.1** lists **typical fouling factors**
More detailed information can be found in TEMA Standards
Fouling information is highly proprietary
$R_0$ plus $R_i$ are commonly assumed to be $= 0.003$ hr ft$^2$ °F/Btu
Calculation of coefficient

**TABLE 1.1** Typical Fouling Factors

| FLUID | $R_0 + R_i$ (HR FT$^2$ °F/BTU) |
| --- | --- |
| Crude oil | 0.002 |
| Natural gas | 0.001 |
| Overhead products | 0.001 |
| Lean oil | 0.002 |
| Rich oil | 0.002 |
| LPG | 0.001 |
| Amine solutions | 0.002 |
| Acid gas | 0.001 |
| Refrigerant liquids | 0.001 |
| Engine exhaust | 0.010 |
| Cooling water | 0.001 |

Often require **trial-and-error** calculations and are **time-consuming** to do by hand

**Computers** are normally **used** to make **detailed calculations**

If calculations are **done by hand,** care should be taken to **ensure** that all **units are consistent**

Reducing film resistances

Most liquids and gases are poor conductors of heat

Film of fluid next to the tube wall generally causes the controlling resistance to heat transfer

Increasing the velocity of the fluid stream can reduce the thickness of the film

Typical U values

Can be used for rough or preliminary calculations

Presented throughout the remainder of this section

## Fouling Considerations

One must not confuse real fouling with a calculation result

Fouling is material accumulation on heat transfer surfaces

The fouling factor is

A calculated result derived from plant data

A fudge factor

A number from a table

Typically one cannot determine how much fouling is shell-side fouling

Evaluating performance

$$R_f = 1/U_{\text{observed}} - 1/U_{\text{Expected}}$$

where

$$1/U_{\text{Expected}} = 1/U_{\text{Clean}} + R_{f(\text{expected})}$$

What is time trend?

What is HEX fouling?

For design
    First $U_{Expected}/U = 125\%$
    Then $R_f = 1/U - 1/U_{Expected}$
    Exceptions
    Will this HEX foul?
    Figure 1.17 illustrates fouling nomenclature

## Fouling Mechanisms
Use trend analysis to help pinpoint the type of fouling
Sources of fouling include
    Particulate or solids deposition
    Salt crystallization or insolubility
    Polymerization, coking, or reaction to solid
    Human additives
    Biological growth
    Corrosion
Determination of fouling causes can be difficult and will
    spotlight the competency of problem-solving tenacity
    Be persistent and keep asking many questions!

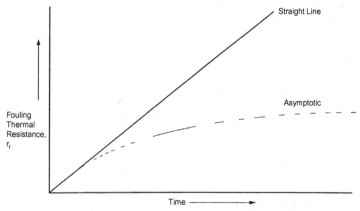

**Figure 1.17** Fouling data trends nomenclature.

**Fouling Prevention**

For particulate or solids deposition

    Maintain sufficient velocity to keep from settling

Salt crystallization or insolubility

    Manage concentration, skin and bulk temperatures, and pH

Polymerization, coking, or reaction to solid

    Keep temperature below activation or

    Control exponential reaction rate

Human additives

    Caution: rather than boil, usually decompose—we recommend avoiding their use

Biological growth

    Materials, temperature, and velocity dependent

    Kill or control rate

Corrosion

    Adjust process to prevent

    Select appropriate material

Caveat: more area rarely allows for operation with active fouling

Figure 1.18 is a graph displaying three different fouling trends one can use to help identify the cause of fouling

### Heat Transfer Research Inc. (HRTI) Computer Simulation Programs

Workhorse for the heat exchanger suite, which

    Is relatively easy to set up

    Used for design and troubleshooting

    Handles many service types, including

        Air coolers

        Sensible heat transfer

        Single phase, except for "two phase" with less than 70% duty as latent heat

        Condensing

        Tube-side and shell-side boiling

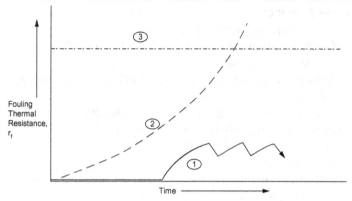

**Figure 1.18** Fouling mechanisms.

Vertical thermosiphons
Horizontal thermosiphons
Kettle reboilers
Plate and frame heat exchangers

Fouling is typically rapid and limits heat transfer dramatically; see Table 1.2.

## Calculating Film Coefficients
Depends on service
Determine level of accuracy
Available tools include:
  Equations
  Charts
  Simplifying assumptions
  Computer programs

**TABLE 1.2** Services Susceptible to Fouling

| PROBLEM | SUSCEPTIBLE SERVICE | AVOID BY |
|---------|---------------------|----------|
| Particulate or solids deposition | Any stream—liquid, gas, or two phase | Maintain sufficient velocity to keep from settling |
| Salt crystallization or insolubility | Stream containing water phase | Manage concentration, skin and bulk temperatures, and pH |
| Polymerization, coking, or reaction to solid | Heavy hydrocarbon liquids and streams contaminated with catalyst | Keep temperature below activation or control exponential reaction rate |
| Man-made additives | Any liquid or two-phase streams | Avoid |
| Biological growth | Water or water-contaminated streams | Kill or control rate of growth. This is materials, temperature, and velocity dependent |
| Corrosion | Practically any stream | Adjust process to prevent corrosion and select appropriate materials to resist corrosion |

## For All but the Simplest Cases, Use a Computer Program

For quicker estimates, use Table 1.3 for approximate film coefficients

For sensible streams, calculating film coefficients is straight-forward

If economic sizing criteria are used, velocity or pressure gradient is optimized and equations can be solved

Economic Sizing Criteria

Equalize or balance film (adjusted) coefficients

For new designs, balance pumping or compression cost against cost of heat transfer surface area

Velocity considerations

0.4 to 0.6 psi/axial ft (9048 to 13,570 Pa/axial m) shell side $\cong$ 2 to 3 ft/s for liquids (0.61 to 0.91 m/s)

**TABLE 1.3** Approximate Film Coefficients

| SERVICE OR FLUID | SHELL OR TUBE SIDE FILM COEFFICIENT BTU/HR °F FT$^2$ (BASED ON BARE OUTSIDE AREA) |
|---|---|
| Sensible | |
| Pure water | 1400 |
| HC, 0.5 cP | 400 |
| HC, 2.0 cP | 250 |
| HC, 10 cP | 150 |
| Gases | |
| Light HC, 150 psig | 100 |
| Air, 10 psig | 15 |
| Air, 300 psig | 60 |
| Condensing | |
| Steam | 1000 |
| Light HC | 200 |
| Heavy HC | 100 |
| Subcooling | 50 |
| Boiling | |
| Water | 1000 |
| Light HC | 300 |
| Heavy HC | 150 |
| Air cooled (fin fan) | |
| Air side | 175 |

0.2 to 0.3 psi/ft (4524 to 6784 Pa/m) tube side [3/4–in. tubes (19.05 mm)[ ≅ 7 to 9 ft/s for liquids (2.1 to 2.7 m/s)

Expensive metallurgy suggests design at a higher velocity

Active fouling (contaminated streams) suggests design at a higher velocity

You must consider a turndown operation and keep the pressure gradient above the particulate fouling threshold

0.1 psi/axial ft (2262 Pa/axial m) shell-side ≅ 1 ft/s (0.305 m/s)

0.05 psi/ft (1131 Pa/m) tube side ≅ 4 ft/s (1.22 m/s)

Use economic sizing criteria for single-phase, two-phase, multicomponent condensing, and flow boiling

Because pure component condensing and boiling typically have very large film coefficients, accuracy is not important

Multicomponent condensing and boiling, as well as a combination of sensible plus latent heat transfer, are very complicated

These should be analyzed using an appropriate computer program

For boiling streams, design and operate in the nucleate regime

For condensing streams, the regime changes dramatically as heat is transferred

See Figure 1.19.


## Inside Film Coefficient

Represents the **resistance to heat flow** caused by the **change** in **flow regime** from

Turbulent flow in the center of the tube

Laminar flow at the tube surface

Calculated from

$$h_i = \left[ 0.022 \left( \frac{dG}{\mu} \right)^{08} \left( \frac{C_p}{k} \right)^{0.4} \left( \frac{\mu}{\mu_w} \right)^{0.16} \right] \frac{k}{p} \qquad (1.27)$$

where

$h_i$ = fluid film heat transfer coefficient, Btu/hr ft$^2$ °F

$D$ = tube inside diameter, ft

$k$ = fluid thermal conductivity, Btu/hr ft °F (refer to Figures 1.20–1.22)

$$G \quad = \quad \text{mass velocity of fluid, lb/hr ft}^2 \text{ [refer to Equations (1.23) and (1.24)]}$$

$$\quad = \quad \text{fluid viscosity, lb/hr ft (2.41) } \mu_{CP}$$

$$\quad = \quad \text{refer to Figures 1.23–1.24}$$

$$\mu_C \quad = \quad \text{fluid viscosity at tube wall, lb/hr ft}$$

$$C_P \quad = \quad \text{fluid specific heat, Btu/lb °F}$$

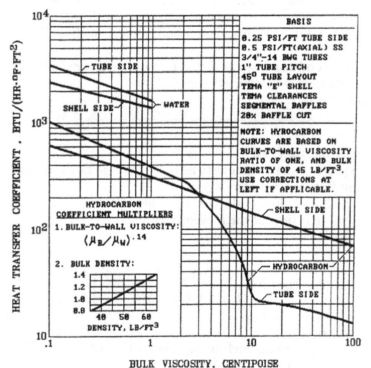

**Figure 1.19** Heat transfer coefficients.

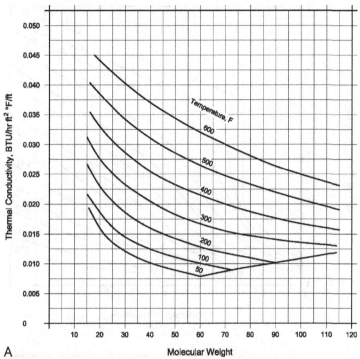

**A**  Molecular Weight

**Figure 1.20A** Thermal conductivity of hydrocarbon gases at standard pressure (14.67 psia) (value is multiplied by the ratio $k/k_A$ from Figure 1.20B).

## Mass Velocity of a Fluid
Liquids

$$G = 18.6 \frac{Q_L\,(SG)}{D^2} \qquad (1.28)$$

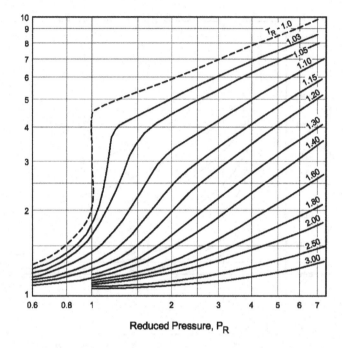

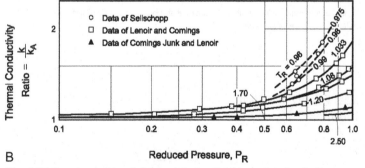

**Figure 1.20B** Thermal conductivity ratio for gases (value is multiplied by the thermal conductivity value from Figure 1.20A).

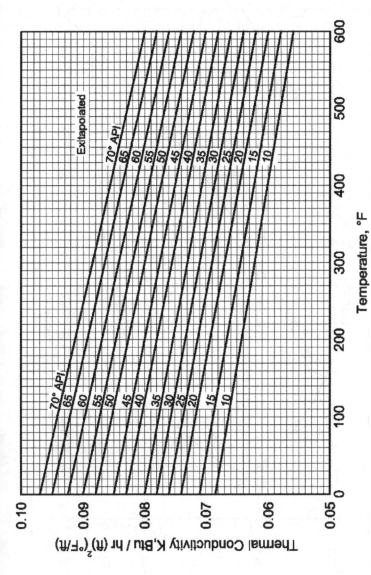

Figure 1.21 Thermal conductivity of hydrocarbon liquids.

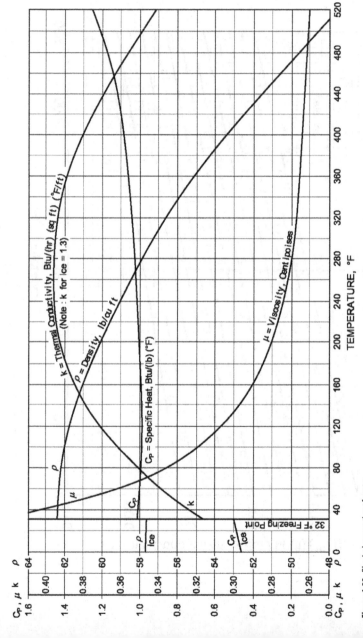

Figure 1.22 Physical properties of water.

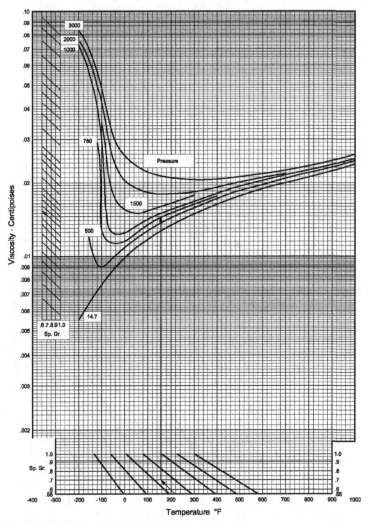

**Figure 1.23** Hydrocarbon gas viscosity.

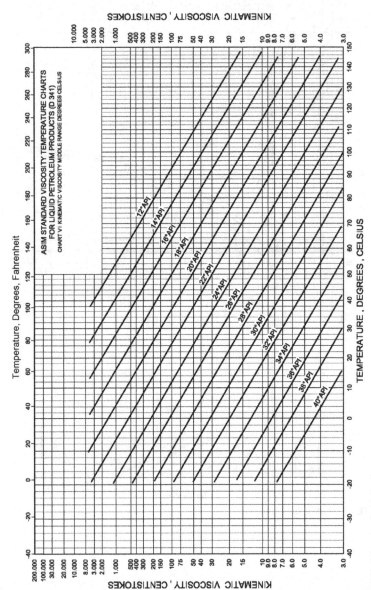

**Figure 1.24** Hydrocarbon liquid viscosity.

Gases

$$G = 4053 \frac{Q_g S}{D^2} \qquad (1.29)$$

where

$Q_L$  =  liquid flow rate per tube, BPD
$Q_g$  =  gas flow rate per tube, MMscfd
$SG$  =  liquid-specific gravity (relative to water)
$S$  =  gas-specific gravity (relative to air)
$D$  =  tube inside diameter, ft

Thermal conductivity, viscosity, and specific heat of various heat medium fluids are given in Figures 1.25–1.29.

### Outside Film Coefficient in a Liquid Bath

Result of natural or free convection

Temperature variations in the fluid cause density variations

Density variations cause the fluid to circulate, which produces free convective heat transfer

For horizontal pipes and tubes spaced more than one diameter apart, the following applies

$$h_0 = 116 \left[ \frac{k^3 C^2 \, \rho \beta \Delta t_m}{\mu d_0} \right]^{0.25} \qquad (1.30)$$

where

$h_0$  =  outside film coefficient, Btu/hr ft$^2$ °F
$k$  =  bath fluid thermal conductivity, Btu/hr ft °F (refer to Table 1.4)
$C$  =  bath fluid heat capacity, Btu/lb °F
$\rho$  =  bath fluid density, lb/ft$^3$
$\mu$  =  bath fluid viscosity, C$_P$
$\Delta t_m$  =  mean temperature difference, °F
$d_0$  =  pipe outside diameter, in.
$\beta$  =  bath fluid coefficient of thermal expansion, 1/°F (Refer to Table 1.5)

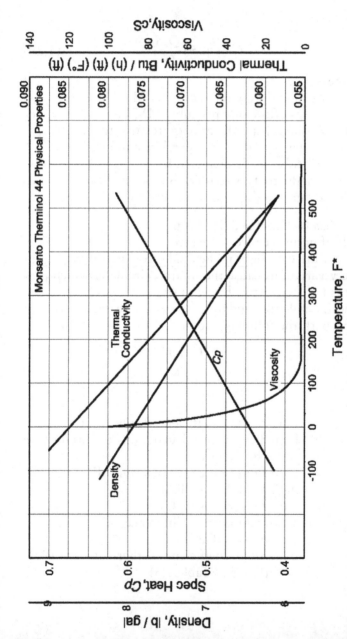

Figure 1.25 Heat transfer coefficient, Monsanto Therminol 44 (use −50° to 450°F; maximum film temperature).

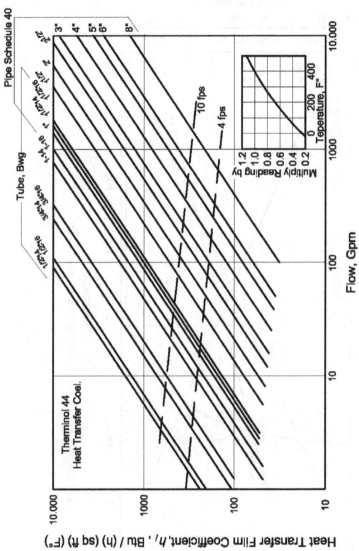

Figure 1.25 *Cont'd*

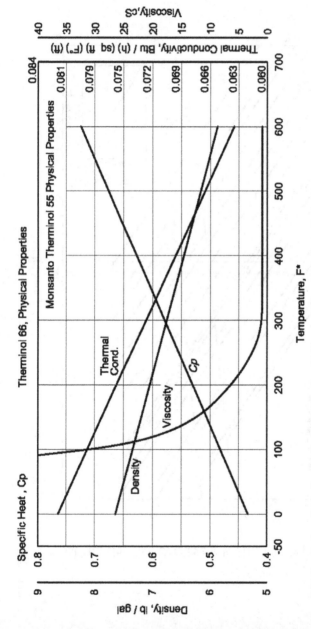

Figure 1.26 Heat transfer coefficient, Monsanto Therminol 55 (use 0° to 600°F; maximum film temperature 635°F).

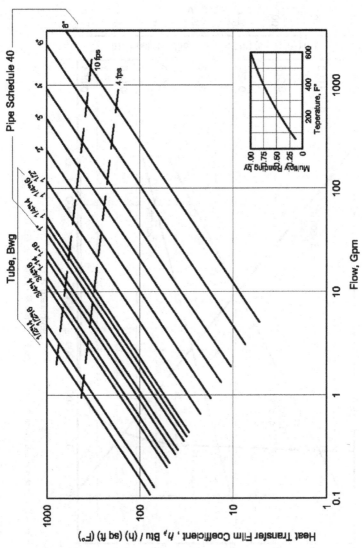

Figure 1.26 *Cont'd*

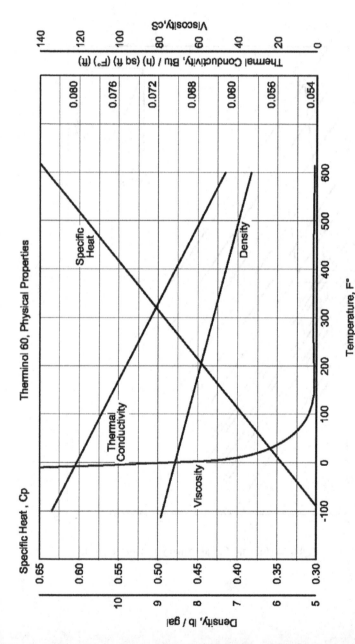

Figure 1.27  Heat transfer coefficient, Monsanto Therminol 55 (use 0° to 600°F; maximum film temperature 635°F).

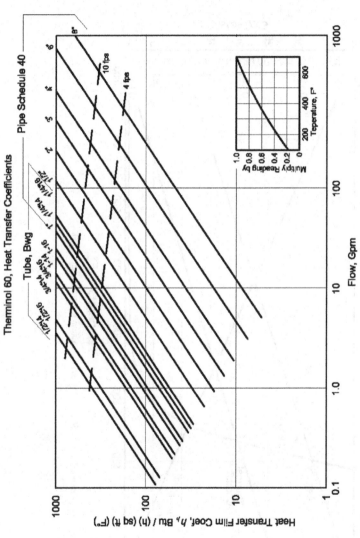

Figure 1.27 *Cont'd*

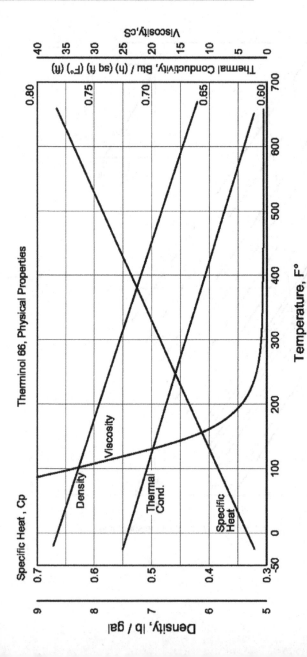

Figure 1.28 Heat transfer coefficient, Monsanto Therminol 66 (use 0° to 650°F; maximum film temperature 750°F).

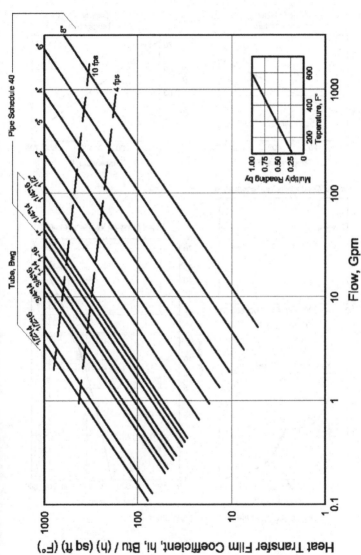

Figure 1.28 *Cont'd*

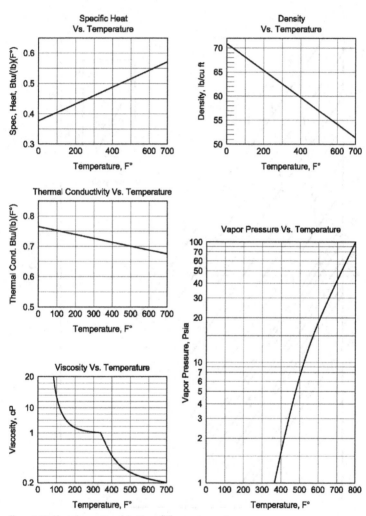

**Figure 1.29** Physical properties of dowtherm "G" versus temperature.

**TABLE 1.4** Thermal Conductivity of Water

| TEMPERATURE (°F) | THERMAL CONDUCTIVITY (BTU/HR FT°F) |
|---|---|
| 32 | 0.343 |
| 100 | 0.363 |
| 200 | 0.393 |
| 300 | 0.395 |
| 420 | 0.376 |
| 620 | 0.275 |

**TABLE 1.5** Coefficients of Thermal Expansion (β)

| BATH FLUID | COEFFICIENT (1/°F) |
|---|---|
| Water | 0.0024 |
| Dowtherms | 0.00043 |
| Therminols | 0.00039 |
| Mobiltherms | 0.0035 |

Bath fluid coefficients of thermal expansion are given in Table 1.5.

Density of water is one divided by the specific volume given in the steam tables (Table 1.6)

### Outside Film Coefficient for Shell-and-Tube Exchangers

For shell and tube heat exchangers with
Shell-side baffles
Shell-side fluid flow perpendicular to tubes
Determined from:

$$h_0 = 0.6K\left(\frac{C\mu_e}{k}\right)^{0.33}\left(\frac{DG_{max}}{\mu_e}\right)^{0.6}\left(\frac{k}{D}\right) \qquad (1.31)$$

where
$h_0$ = outside film coefficient, Btu/hr ft$^2$ °F
$D$ = tube outside diameter, ft

**TABLE 1.6** Properties of Dry Saturated Steam

| TIME θ | PRESS psig | SPECIFIC VOLUME | | | ENTHALPY | | | ENTROPY | | | TIME θ |
|---|---|---|---|---|---|---|---|---|---|---|---|
| TEMP °F T | P | SATURATED LIQUID Vliq | RVAP. | SATURATED VAPOR Vvap | SATURATED VAPOR Hliq | RVAP. | SATURATED VAPOR hvap | SATURATED LIQUID heliq | RVAP. | SATURATED VAPOR hevap | TEMP °F T |
| 32 | 0.08854 | 0.01602 | 3308 | 3208 | 0.00 | 1075.3 | 1075.8 | 0.0000 | 2.1877 | 2.1877 | 32 |
| 35 | 0.09995 | 0.01602 | 2947 | 2947 | 3.02 | 1074.1 | 1077.1 | 0.0061 | 2.1709 | 2.1770 | 35 |
| 40 | 0.12170 | 0.01602 | 2444 | 2444 | 8.05 | 1071.3 | 1079.3 | 0.0162 | 2.1435 | 2.1597 | 40 |
| 45 | 0.14752 | 0.01602 | 2036.4 | 2036.4 | 13.06 | 1068.4 | 1081.5 | 0.0262 | 2.1167 | 2.1429 | 45 |
| 50 | 0.17811 | 0.01603 | 1703.2 | 1703.2 | 18.07 | 1065.8 | 1083.7 | 0.0361 | 2.0903 | 2.1264 | 50 |
| 60 | 0.2583 | 0.01604 | 1206.8 | 1206.7 | 28.06 | 1059.9 | 1088.0 | 0.0555 | 2.0393 | 2.0948 | 60 |
| 70 | 0.3631 | 0.01605 | 867.8 | 857.9 | 38.04 | 1054.4 | 1092.3 | 0.0745 | 1.9902 | 2.0647 | 70 |
| 80 | 0.5069 | 0.01608 | 632.1 | 633.1 | 48.02 | 1048.6 | 1096.6 | 0.0932 | 1.9428 | 2.0380 | 80 |
| 90 | 0.5982 | 0.01610 | 468.0 | 468.0 | 57.99 | 1042.9 | 1100.9 | 0.1115 | 1.8972 | 2.0087 | 90 |
| 100 | 0.9492 | 0.01613 | 350.3 | 350.4 | 67.97 | 1037.2 | 1105.2 | 0.1295 | 1.8531 | 1.9828 | 100 |
| 110 | 1.2748 | 0.01617 | 265.3 | 265.4 | 77.94 | 1031.8 | 1109.5 | 0.1471 | 1.8106 | 1.9577 | 110 |
| 120 | 1.6924 | 0.01620 | 203.25 | 203.27 | 87.92 | 1025.8 | 1113.7 | 0.1645 | 1.7694 | 1.9339 | 120 |
| 130 | 2.2225 | 0.01625 | 157.32 | 157.34 | 97.90 | 1020.0 | 1117.9 | 0.1818 | 1.7256 | 1.9112 | 130 |
| 140 | 2.2886 | 0.01629 | 122.99 | 123.01 | 107.39 | 1014.1 | 1122.0 | 0.1984 | 1.6910 | 1.3894 | 140 |
| 150 | 3.718 | 0.01634 | 97.04 | 97.07 | 117.99 | 1008.2 | 1126.1 | 0.2149 | 1.6537 | 1.3685 | 150 |
| 160 | 4.741 | 0.01629 | 77.27 | 77.29 | 127.89 | 1002.3 | 1130.2 | 0.2311 | 1.6174 | 1.3485 | 160 |
| 170 | 5.992 | 0.01615 | 62.04 | 62.06 | 137.93 | 996.3 | 1134.2 | 0.2472 | 1.5822 | 1.8293 | 170 |
| 180 | 7.510 | 0.01651 | 50.21 | 50.23 | 147.92 | 990.2 | 1138.1 | 0.2630 | 1.5480 | 1.8109 | 180 |
| 190 | 9.339 | 0.01657 | 40.94 | 40.96 | 157.95 | 984.1 | 1142.0 | 0.2785 | 1.5147 | 1.7932 | 190 |
| 200 | 11.528 | 0.01663 | 33.62 | 33.64 | 167.99 | 977.9 | 1145.9 | 0.2938 | 1.4824 | 1.7762 | 200 |

| | | | | | | | | | | | |
|---|---|---|---|---|---|---|---|---|---|---|---|
| 210 | 14.123 | 0.01670 | 27.80 | 27.32 | 178.05 | 971.6 | 1149.7 | 0.3090 | 1.4508 | 1.7398 | 210 |
| 212 | 14.696 | 0.01672 | 26.73 | 26.80 | 180.07 | 970.3 | 1150.4 | 0.3120 | 1.4446 | 1.7568 | 212 |
| 220 | 17.186 | 0.01677 | 23.13 | 22.15 | 188.13 | 965.2 | 1153.4 | 0.3239 | 1.4201 | 1.7440 | 220 |
| 230 | 20.780 | 0.01684 | 19.265 | 19.382 | 198.23 | 958.8 | 1157.0 | 0.3387 | 1.3901 | 7238 | 230 |
| 240 | 24.969 | 0.01692 | 16.208 | 16.223 | 208.34 | 952.2 | 1160.5 | 0.3531 | 1.3609 | 1.7140 | 240 |
| 250 | 29.823 | 0.01700 | 13.804 | 13.821 | 216.48 | 945.5 | 1164.0 | 0.3875 | 1.3323 | 1.6998 | 250 |
| 260 | 35.429 | 0.01709 | 11.746 | 11.763 | 225.64 | 938.7 | 1167.3 | 0.3817 | 1.3043 | 1.6860 | 260 |
| 270 | 41.858 | 0.01717 | 10.044 | 10.061 | 238.84 | 931.3 | 1170.6 | 0.3958 | 1.2769 | 1.6727 | 270 |
| 280 | 49.203 | 0.01726 | 8.628 | 8.645 | 249.06 | 924.7 | 1173.8 | 0.4098 | 1.2301 | 1.6597 | 280 |
| 290 | 37.558 | 0.01735 | 7.444 | 7.481 | 259.31 | 917.5 | 1176.8 | 0.4234 | 1.2238 | 1.6472 | 290 |
| 300 | 67.013 | 0.01745 | 6.449 | 6.468 | 289.59 | 910.1 | 1179.7 | 0.4399 | 1.1980 | 1.8330 | 300 |
| 310 | 77.68 | 0.01755 | 5.609 | 5.523 | 279.92 | 902.6 | 1182.5 | 0.4504 | 1.1727 | 1.6231 | 310 |
| 320 | 89.68 | 0.01765 | 4.896 | 4.914 | 290.23 | 894.9 | 1185.2 | 0.4637 | 1.1478 | 1.6115 | 320 |
| 330 | 103.06 | 0.01778 | 4.289 | 4.307 | 300.68 | 887.0 | 1187.7 | 0.4760 | 1.1233 | 1.6002 | 330 |
| 340 | 118.01 | 0.01787 | 3.770 | 3.788 | 311.13 | 879.0 | 1190.1 | 0.4900 | 1.0992 | 1.5891 | 340 |
| 350 | 134.63 | 0.01799 | 3.324 | 3.342 | 321.53 | 870.7 | 1192.3 | 0.5029 | 1.0754 | 1.5783 | 350 |
| 360 | 153.04 | 0.01811 | 2.939 | 2.957 | 332.18 | 862.2 | 1194.4 | 0.5158 | 1.0519 | 1.3677 | 350 |
| 370 | 173.37 | 0.01823 | 2.606 | 2.625 | 342.79 | 853.5 | 1196.3 | 0.5288 | 1.0287 | 1.5573 | 370 |
| 380 | 195.77 | 0.01826 | 2.317 | 2.335 | 353.45 | 844.8 | 1198.1 | 0.5413 | 1.0059 | 1.5471 | 380 |
| 390 | 220.37 | 0.01850 | 2.0651 | 2.0838 | 364.17 | 835.4 | 1199.6 | 0.3539 | 0.9832 | 1.5371 | 390 |
| 400 | 247.31 | 0.01864 | 1.3447 | 1.8633 | 374.97 | 328.0 | 1201.0 | 0.5684 | 0.9608 | 1.5272 | 400 |
| 410 | 278.75 | 0.01878 | 1.6512 | 1.6700 | 385.53 | 816.3 | 1202.1 | 0.5788 | 0.9388 | 1.5174 | 410 |
| 420 | 308.83 | 0.01894 | 1.4811 | 1.5000 | 396.77 | 806.3 | 1203.1 | 0.5912 | 0.9168 | 1.5078 | 420 |
| 430 | 343.72 | 0.01910 | 1.3308 | 1.3499 | 407.79 | 796.0 | 1203.8 | 0.6035 | 0.8947 | 1.4982 | 430 |
| 440 | 281.59 | 0.01828 | 1.1979 | 1.2171 | 418.90 | 785.4 | 1204.3 | 0.6158 | 0.8730 | 1.4887 | 440 |

(Continued)

**TABLE 1.6** Properties of Dry Saturated Steam—*cont'd*

| TEMP °F T | PRESS psig P, p | SPECIFIC VOLUME SATURATED LIQUID Vliq | SPECIFIC VOLUME RVAP. | SPECIFIC VOLUME SATURATED VAPOR Vvap | ENTHALPY SATURATED VAPOR Hliq | ENTHALPY RVAP. | ENTHALPY SATURATED VAPOR hvap | ENTROPY SATURATED LIQUID heliq | ENTROPY RVAP. | ENTROPY SATURATED VAPOR hevap | TEMP °F T |
|---|---|---|---|---|---|---|---|---|---|---|---|
| 450 | 422.8 | 0.0194 | 1.0799 | 1.0993 | 430.1 | 774.5 | 1204.8 | 0.6290 | 0.3513 | 1.4793 | 450 |
| 460 | 466.9 | 0.0196 | 0.9748 | 0.9944 | 441.4 | 783.2 | 1204.8 | 0.6402 | 0.8298 | 1.4700 | 460 |
| 470 | 514.7 | 0.0198 | 0.8811 | 0.9009 | 452.5 | 751.3 | 1204.3 | 0.6523 | 0.3083 | 1.4806 | 470 |
| 480 | 566.1 | 0.0200 | 0.7972 | 0.8172 | 464.4 | 739.4 | 1203.7 | 0.6645 | 0.7868 | 1.4513 | 480 |
| 490 | 621.4 | 0.0202 | 0.7221 | 0.7423 | 476.0 | 726.8 | 1202.8 | 0.6768 | 0.7653 | 1.4419 | 490 |
| 500 | 680.8 | 0.0204 | 0.6545 | 0.6749 | 487.8 | 713.0 | 1201.7 | 0.5887 | 0.7438 | 1.4323 | 500 |
| 520 | 812.4 | 0.0209 | 0.5385 | 0.5594 | 511.9 | 688.4 | 1198.2 | 0.7130 | 0.7006 | 1.4138 | 520 |
| 540 | 962.5 | 0.0215 | 0.4434 | 0.4549 | 538.8 | 656.6 | 1193.2 | 0.7374 | 0.6568 | 1.2942 | 540 |
| 560 | 1133.1 | 0.0221 | 0.2647 | 0.3868 | 582.2 | 624.2 | 1186.4 | 0.7821 | 0.6121 | 1.3742 | 560 |
| 580 | 1325.2 | 0.0228 | 0.2988 | 0.2217 | 588.9 | 588.4 | 1177.3 | 0.7872 | 0.5659 | 1.2532 | 580 |
| 600 | 1542.9 | 0.0228 | 0.2432 | 0.2668 | 617.0 | 548.5 | 1165.3 | 0.8131 | 0.5178 | 1.3307 | 600 |
| 620 | 1738.5 | 0.0247 | 0.1958 | 0.2201 | 646.7 | 503.8 | 1150.3 | 0.8398 | 0.4664 | 1.2062 | 620 |
| 640 | 2059.7 | 0.0250 | 0.1538 | 0.1798 | 678.6 | 452.0 | 1130.3 | 0.8579 | 0.4110 | 1.2789 | 640 |
| 660 | 2265.4 | 0.0275 | 0.1165 | 0.1442 | 714.2 | 390.2 | 1104.4 | 0.8987 | 0.3485 | 1.2472 | 660 |
| 680 | 2708.1 | 0.0305 | 0.0510 | 0.1115 | 757.3 | 309.9 | 1067.2 | 0.9351 | 0.2219 | 1.2071 | 680 |
| 700 | 2093.7 | 0.0369 | 0.0392 | 0.0761 | 823.3 | 172.1 | 995.4 | 0.9905 | 0.1484 | 1.1389 | 700 |
| 705.4 | 3206.2 | 0.0503 | 0 | 0.0503 | 902.7 | 0 | 902.7 | 1.0580 | 0 | 1.0580 | 705.4 |

DENSITY OF WATER = 1 ÷ SPECIFIC VOLUME

LATENT HEAT = $h_{fg}$

$k$ = fluid thermal conductivity, Btu/hr ft °F
$G_{max}$ = maximum mass velocity of fluid, lb/hr ft$^2$
$C$ = fluid specific heat, Btu/lb °F
$\mu_e$ = fluid viscosity, lb/hr ft
$K$ = coefficient from Table 1.7

**Tube Metal Resistant**

$$r_w = \frac{\Delta X_w}{K_w} = \frac{L}{K}$$
(1.32)

where
$\Delta X_w$ = wall thickness, ft
$K_w$ = thermal conductivity of pipe, Btu/hr ft °F (Table 1.10)

Tables 1.8 and 1.9 have basic tube and coil properties for use in Equation (1.10), and Table 1.10 lists the thermal conductivity of different metals at 200°F.

**Approximate Overall Heat Transfer Coefficients**

General considerations

Calculation of "U" using the previous equation is tedious
Specialists use computer programs to calculate this value

**TABLE 1.7** "$K$" Value for Fluid Flow Perpendicular to a Bank of Staggered Tubes "$N$" Rows Deep

| $N$ | $K$ |
|---|---|
| 1 | 0.24 |
| 2 | 0.27 |
| 3 | 0.29 |
| 4 | 0.30 |
| 5 | 0.31 |
| 6 | 0.32 |
| 10 | 0.33 |

**TABLE 1.8** Characteristics of Tubing

| TUBE OD IN. | B.W.G GAUGE | THICKNESS IN. | INTERNAL AREA IN. | FT EXTERNAL SURFACE PER FT LENGTH | FT INTERNAL SURFACE PER FT LENGTH |
|---|---|---|---|---|---|
| 1 | 14 | .083 | .5463 | .2618 | .2183 |
| 1 | 15 | .072 | .5755 | .2618 | .2241 |
| 1 | 16 | .065 | .5945 | .2618 | .2278 |
| 1 | 18 | .049 | .6390 | .2618 | .2361 |
| 1 | 20 | .035 | .6793 | .2618 | .2435 |
| 11/4 | 7 | .180 | .6221 | .3272 | .2330 |
| 11/4 | 8 | .165 | .6648 | .3272 | .2409 |
| 11/4 | 10 | .134 | .7574 | .3272 | .2571 |
| 11/4 | 11 | .120 | .8012 | .3272 | .2644 |
| 11/4 | 12 | .109 | .8365 | .3272 | .2702 |
| 11/4 | 13 | .095 | .8825 | .3272 | .2775 |
| 11/4 | 14 | .083 | .9229 | .3272 | .2838 |
| 11/4 | 16 | .065 | .9852 | .3272 | .2932 |
| 11/4 | 18 | .049 | 1.042 | .3272 | .3016 |
| 11/4 | 20 | .035 | 1.094 | .3272 | .3089 |
| 11/2 | 10 | .134 | 1.192 | .3927 | .3225 |
| 11/2 | 12 | .109 | 1.291 | .3927 | .3356 |
| 11/2 | 14 | .083 | 1.398 | .3927 | .3492 |
| 11/2 | 16 | .065 | 1.474 | .3927 | .3587 |
| 2 | 11 | .120 | 2.433 | .5236 | .4608 |
| 2 | 12 | .109 | 2.494 | .5236 | .4665 |
| 2 | 13 | .095 | 2.573 | .5236 | .4739 |
| 2 | 14 | .083 | 2.642 | .5236 | .4801 |

**TABLE 1.9** Pipe Coil Data

| NOM. SIZE IN. | SCH. NO. | OD IN. | ID IN. | INTERNAL SURFACE AREA (FT²/FT) | EXTERNAL SURFACE AREA (FT²/FT) |
|---|---|---|---|---|---|
| 1 | S40 | 1.315 | 1.049 | 0.275 | 0.344 |
| | X80 | | 0.957 | 0.251 | |
| | 160 | | 0.815 | 0.213 | |
| | XX | | 0.599 | 0.157 | |
| 2 | S40 | 2.375 | 2.067 | 0.541 | 0.622 |
| | X80 | | 1.939 | 0.508 | |
| | 160 | | 1.687 | 0.442 | |
| | XX | | 1.503 | 0.394 | |
| 21/2 | XXX | 2.875 | 1.375 | 0.360 | 0.753 |
| 3 | S40 | 3.50 | 3.068 | 0.803 | 0.916 |
| | X80 | | 2.900 | 0.759 | |
| | 160 | | 2.624 | 0.687 | |
| | XX | | 2.300 | 0.602 | |
| 4 | S40 | 4.50 | 4.026 | 1.054 | 1.19 |
| | X80 | | 3.826 | 1.002 | |
| | 160 | | 3.438 | 0.900 | |
| | XX | | 3.152 | 0.825 | |

**TABLE 1.10** Thermal Conductivity of Metals at 200°F

| MATERIAL | CONDUCTIVITY BTU/(HR FT °F) |
|---|---|
| Copper | 223 |
| Brass (admiralty) | 70 |
| Silicon bronze | 15 |
| Stainless steel (18cr-8ni) | 8 |
| Inconel | 8 |
| 90-10 CuNi | 30 |
| 70-30 | 18 |
| Monel | 15 |
| Titanium | 10 |

"Quick look" tables from GPSA Engineering Data Book (Tables 1.11 and 1.12)

Provides approximate values of "U" for shell-and-tube heat exchangers

Exchanging water with 100 psi gas gives a low U value, thus requiring a high surface area

Exchanging water with 1000 psi gas gives a much higher U value, thus requiring less surface area in the exchanger

If water is being exchanged with water, a very high U value is achieved

Values in the table do not differentiate between tube-side and shell-side fluids

Placement of fluid does make a difference to the U value

"Quick Look" graphs from Smith Industries

Figure 1.30 provides approximate values of "U" for exchange from water bath to a natural gas stream in a coil

**TABLE 1.11** Typical Bare Tube Overall Heat Transfer Coefficients, for Shell-and-Tube Heat Exchangers (Btu/hr ft$^2$ °F)

| SERVICE | U | SERVICE | U |
|---|---|---|---|
| Water with 100 psi | 35–40 | Wtr condensers with $C_3$, $C_4$ | 125–135 |
| Gas water with 300 psi | 40–50 | Wtr condensers with naphtha | 70–80 |
| Gas water with 700 psi | 60–70 | Wtr condensers with still ovhd | 70–80 |
| Gas water with 1000 psi | 80–100 | Wtr condensers with amine | 100–110 |
| Gas water with kerosene | 80–90 | Reboilers w/steam | 140–160 |
| Water with MEA | 130–150 | Reboilers w/hot oil | 90–120 |
| Water with air | 20–25 | 100 psi gas w/500 psi gas | 50–70 |
| Water with water | 180–200 | 1000 psi gas w/1000 psi gas | 60–80 |
| Oil with oil | 80–100 | 1000 psi gas chiller (gas-$C_3$) | 60–80 |
| $C_3$ with $C_3$ liquid | 110–130 | MEA exchanger | 120–130 |

*Notes:* Maximum boiling film transfer coefficient. Hydrocarbons: 300 to 500 Btu/hr ft$^2$°F. Water: 2000 Btu/hr ft$^2$ °F.

Figure 1.31 is a nomograph for crude oil streams heated with a water bath

**Summary**
Shell-and-tube heat exchanger sizing
  Calculate **MTD**
  Calculate **"U"**
    Use heat transfer equations
    Use **GPSA procedure** in "Engineering Data Book"
    Use **approximate overall heat transfer coefficient** from tables, nomographs, or graph

**TABLE 1.12** Typical "U" Value Ranges for Shell-and-Tube Exchangers

| SERVICE | U, BTU/(HR FT$^2$ °F) |
|---|---|
| Water coolers | |
| Gas (to 500 psi) | 35–50 |
| Gas (500 to 1000 psi) | 50–80 |
| Gas (over 1000 psi) | 80–100 |
| Natural gasoline | 70–90 |
| MEA | 130–150 |
| Air | 15–25 |
| Water | 170–200 |
| Water condensers | |
| Amine regenerator | 100–110 |
| Fractionator overhead | 70–80 |
| Light hydrocarbons | 85–135 |
| Reboilers | |
| Steam | 140–160 |
| Hot oil | 90–120 |
| Glycol | 10–20 |
| Amine | 100–120 |
| General | |
| Oil–oil | 80–100 |
| Propane–propane | 100–130 |
| Rich MEA–lean MEA | 120–130 |
| Gas–gas (to 500 psi) | 50–70 |
| Gas–gas (about 10,000 psi) | 55–75 |
| Gas–propane chiller | 60–90 |

## ▶ DETERMINATION OF PROCESS HEAT DUTY

### Overview

Heat required to be added or be removed from the process fluids to create the required change in temperature

Can be in the form of sensible heat, latent heat, or latent heat of vaporization

### Sensible Heat

Amount of **heat absorbed or lost** by a substance that **causes a change** in the **temperature** of the substance

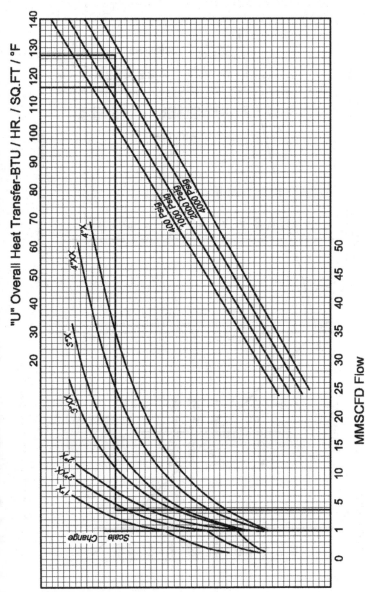

Figure 1.30 Values of U for exchange from water bath to natural gas stream in a coil.

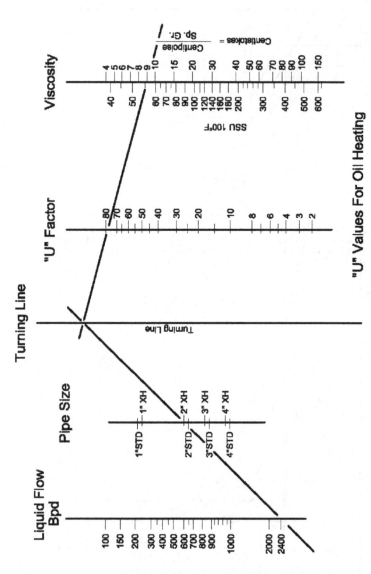

Figure 1.31   Values of U for crude oil streams with water bath heaters.

Example: as **heat is added to steel,** the **temperature** of the
**steel increases** and can be **measured**
**General equation**

$$q_{sh} = WC_p\Delta T \tag{1.33}$$

where
$q_{sh}$  =  sensible heat duty, Btu/hr (kcal/hr)
$W$  =  mass flow rate, lb/hr (kg/hr)
$C_p$  =  specific heat of the fluid, Btu/lb-°F(kcal/kg-°C)
  =  Figures 1.32 and 1.33
$\Delta T$  =  temperature change for a stream °F (°C)

**Latent Heat**
When a substance changes from a **solid to a liquid** or from a
**liquid to a vapor,** the **heat absorbed** is in the form of **latent
heat**
Amount of heat energy **absorbed** or **lost** by a substance
when changing phases
This heat energy **cannot be sensed** by measuring **the
temperature**
General equation

$$q_{ih} = W\lambda \tag{1.34}$$

where
$q_{ih}$  =  latent heat duty, Btu/hr
$W$  =  mass flow rate, lb/hr
$\lambda$  =  latent heat, Btu/lb
Latent heat of vaporization for **hydrocarbon compounds** is
given in Table 1.13.
Latent heat of vaporization of **water** is given by $h_{fg}$ **in the
steam tables**

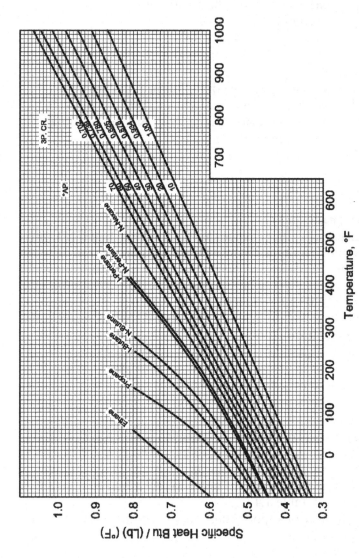

**Figure 1.32** Specific heats of hydrocarbon liquids.

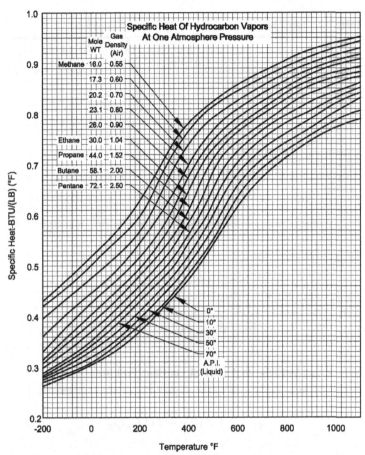

**Figure 1.33** Specific heats of hydrocarbon vapors.

## Heat Duty for Multiphase Streams

Process heat duty when more than one phase exists in the process stream can be calculated using the following equation:

$$q_p = q_g + q_o + q_w \qquad (1.35)$$

**TABLE 1.13** Latent Heat of Vaporization for Hydrocarbon Compounds

| COMPOUND | HEAT OF VAPORIZATION 14.696 PSIA AT BOILING POINT, BTU/LB |
|---|---|
| Methane | 219.22 |
| Ethane | 210.41 |
| Propane | 183.05 |
| n-Butane | 165.65 |
| Isobutane | 157.53 |
| n-Pentane | 153.59 |
| Isopentane | 147.13 |
| Hexane | 143.95 |
| Heptane | 136.01 |
| Octane | 129.53 |
| Noane | 123.76 |
| Decane | 118.68 |

where

$q_p$ = overall heat duty, Btu/hr
$q_g$ = gas heat duty, Btu/hr
$q_o$ = oil heat duty, Btu/hr
$q_w$ = water heat duty, Btu/hr

**Natural Gas Sensible Heat Duty at Constant Pressure**

Sensible heat duty for natural gas at constant pressure is

$$q_g = 41.7 \, (T_2 - T_1) \, C_g Q_g \qquad (1.36)$$

where

$T_2$ = inlet temperature, °F
$T_2$ = outlet temperature, °F
$C_g$ = gas heat capacity, Btu/Mscf °F
$Q_g$ = gas flow rate, MMscfd

Gas heat capacity, $C_g$, is determined at atmospheric conditions and then corrected for temperature and pressure based on reduced pressure and temperature

**Gas Heat Capacity**

General equation

$$C_g = 2.64[29.SC_p + \Delta C_p] \qquad (1.37)$$

where

$C$ = gas-specific heat at one atmosphere pressure, Btu/lb °F

$\Delta C_p$ = correction factor (from Figure 1.34)

$S$ = gas-specific gravity

Correction factor, $\Delta C_p$, is obtained from **Figure 1.34** when the values of $P_r$ and $T_r$ are known

$$T_r = T_a/T_c \qquad (1.38)$$

$$P_r = P/P_C \qquad (1.39)$$

where

$P_r$ = gas reduced pressure

$P$ = gas pressure, psia

$P_c$ = gas pseudo-critical pressure, psia

= Figure 1.35

$T_r$ = gas reduced temperature

$T_a$ = gas average temperature, °R

= $^1/_2(T_1 + T_2)$

$T_c$ = gas pseudo critical temperature, °R

= **Figure 1.35**

**Calculation of Gas Pseudo-Critical Pressure and Temperature**

Approximated from Figure 1.35.

Calculated as a weighted average of critical temperature and pressure of various components on a mole fraction basis

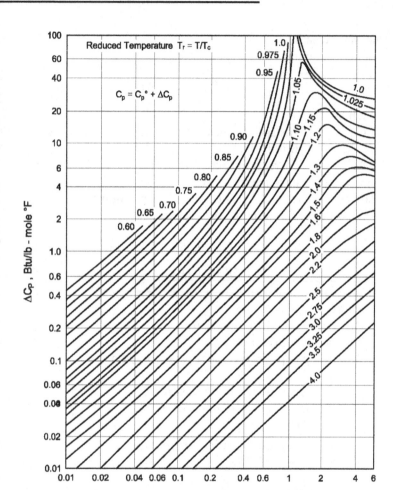

**Figure 1.34** Heat capacity correction factor.

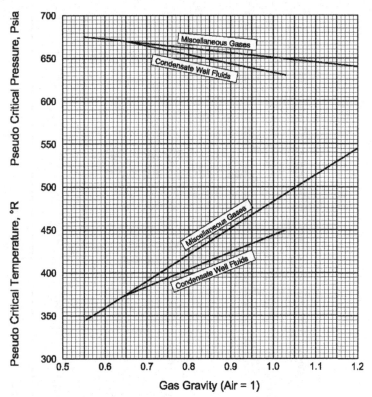

**Figure 1.35** Pseudo-critical properties of natural gas.

(Refer to Table 1.14 for sample calculations)
For greater precision, a correction for $H_2S$ and $CO_2$ content
   should be made

### Oil Sensible Heat Duty
The **sensible heat duty** for the oil phase is:

$$q_0 = 14.6\,(SG)(T_2 - T_1)\,C_0 Q_0 \qquad (1.40)$$

**TABLE 1.14** Estimate of Specific Gravity; Pseudo–Critical Temperature and Pseudo–Critical Pressure, for Example, Field

| COMPONENT | A MOLE PERCENT GAS COMPOSITION | B MOLECULAR WEIGHT | C CRITICAL TEMP °R | D CRITICAL PSIA |
|---|---|---|---|---|
| $CO_2$ | 4.03 | 44.010 | 547.87 | 1071.0 |
| $N_2$ | 1.44 | 28.013 | 227.3 | 493.0 |
| $H_2S$ | 0.0019 | 34.076 | 672.6 | 1036.0 |
| $C_1$ | 85.55 | 16.043 | 343.37 | 667.8 |
| $C_2$ | 5.74 | 30.070 | 550.09 | 707.0 |
| $C_3$ | 1.79 | 44.097 | 666.01 | 616.3 |
| $iC_4$ | 0.41 | 58.124 | 734.98 | 529.1 |
| $nC_4$ | 0.41 | 58.124 | 765.65 | 550.7 |
| $iC_5$ | 0.20 | 72.151 | 829.10 | 490.4 |
| $nC_5$ | 0.13 | 72.151 | 845.70 | 488.6 |
| $C_6$ | 0.15 | 86.178 | 913.70 | 436.9 |
| $C_{7+}$ | 0.15 | 147.0 | 1112.0 | 304.0 |
| Computed value | 100.00 | 19.48 | 374.55 | 680.33 |
| Computation | $\sum A_i$ | $\sum[(A_iB_i)/\sum A_i$ | $\sum[(A_iC_i)/\sum A_i$ | $\sum[(A_iD_i)/\sum A_i$ |
| Specific gravity | $= 19.48/29$ $= 0.67$ | | | |

where
$C_0$ = oil-specific heat, Btu/lb °F (Figure 1.35)
$Q_0$ = oil flow rate, BPD
$SG$ = oil-specific gravity (water = 1)

## Water Sensible Heat Duty

The duty for **heating free water** may be **determined** from the **following equation** assuming a water heat capacity of 1.0 Btu/lb °F

$$q_w = 14.6\,(T_2 - T_1)Q_w \qquad (1.41)$$

where
$Q_w$ = water flow rate, bwpd

## Heat Duty Where There Are Phase Changes

Best to perform flash calculations so as to determine the heat loss or gain by the change in enthalpy
For a "quick look" hand calculation
   Calculate the sensible heat for both the gas and the liquid phases of each component
Sum of all latent and sensible heats is the approximate total heat duty

## Heat Loss to Atmosphere

Calculated using the general equation

$$q = UA\Delta t_m \qquad (1.10)$$

"U" can be calculated from a modification of Equation (1.25). The modification is based on the following

Inside film coefficient is very large compared to the outside film coefficient

Adds a factor for conduction losses through insulation; allows fouling factors to be conservative

Under these conditions, Equation (1.25) reduces to:

$$\frac{1}{U} = \frac{1}{h_0} + \frac{\Delta X_1}{K_1} + \frac{\Delta X_2}{K_2} \qquad (1.42)$$

where

| | | |
|---|---|---|
| $h_0$ | = | $1 + 0.22\, V_w$ ($V_W < 16$ ft/s) |
| | = | $0.53\, V_w^{0.8f}$ ($V_w > 16$ ft/s) |
| $V_w$ | = | wind velocity (ft/s) |
| | = | $1.47 \times$ (mph) |
| $\Delta X_1$ | = | shell thickness, ft |
| $K_1$ | = | shell thermal conductivity, Btu/hr ft °F |
| | = | 30 for carbon steel (Table 1.11) |
| $\Delta X_2$ | = | insulation thickness, ft |
| $K_2$ | = | insulation thermal conductivity, Btu/hr ft °F |
| | = | 0.03 for mineral wool |

For preliminary calculations we often assume

5–10% of $q$ for uninsulated equipment

1–2% of $q$ for insulated vessels

### Heat Transfer from a Fire Tube

Fire tube contains a **flame** burning **inside** a piece of **pipe,** which is in turn **surrounded** by the **process fluid**

Radiant and convective heat transfer from the flame to the inside surface of the fire tube

Conductive heat transfer through the wall thickness of the tube

Convective heat transfer from the outside surface of that tube to the oil being treated

Difficult to solve for the heat transfer in terms of an **overall heat transfer** coefficient

Rather, **what is done** is to **size** the **fire tube** by using a **heat flux rate**

**Heat flux rate** represents the amount of **heat** that can be **transferred** from the **fire tube** of the **process** per **unit area** of **outside surface** of the **fire tube**

**Common heat flux rates** are given in Table 1.15

**Required fire tube area is given by**

$$(Surface\ Area\ of\ Fire\ Tube) = \frac{Heat\ Duty\ Including\ Losses\ (Btu/hr)}{Design\ Flux\ Rate,\ (Btu/hr - ft^2)}$$

$$(1.43)$$

**Example 3:** If total heat duty (sensible heat, latent heat duty, heat losses to atmosphere) was 1 MMBtu/hr and water was being heated, a heat flux rate of 10,000 Btu/hr $ft^2$ would be used yielding 100 $ft^2$ of fire tube area would be required

### Natural Draft Fire Tubes

Minimum **cross-sectional area** of the fire tube is set by limiting the heat release density to 21,000 Btu/hr in.$^2$

**TABLE 1.15** Common Heat Flux Rates

| MEDIUM BEING HEATED | DESIGN HEAT FLUX RATE BTU/HR FT$^2$ |
|---|---|
| Water | 10,000 |
| Boiling water | 10,000 |
| Crude oil | 8,000 |
| Heating medium oils | 8,000 |
| Glycol | 7,500 |
| Amine | 7,500 |

At heat release densities **above** this value, the **flame** may become **unstable** because of **insufficient air**

Using this limit, a **minimum fire tube diameter** is established by

$$d^2 = \frac{Burner\ Heat\ Release\ Density(Btu/hr)}{16,500} \qquad (1.44)$$

where

$d$ = minimum fire tube diameter, in.

Note that the **burner heat release density** will be somewhat **higher** than the **heat duty** (including losses) because a **standard burner size** will be chosen **slightly larger** than that **required**

**Standard burner sizes** and **minimum fire tube diameters** are included in Table 1.16.

Figure 1.36 approximates the combustion efficiency of natural gas (1050 Btu/scf) in emulsion treaters

**TABLE 1.16** Standard Burner Sizes and Minimum Diameter

| BTU/HR | MINIMUM DIAMETER, IN. |
|--------|-----------------------|
| 100,000 | 2.5 |
| 250,000 | 3.9 |
| 500,000 | 5.5 |
| 750,000 | 6.7 |
| 1,000,000 | 7.8 |
| 1,500,000 | 9.5 |
| 2,000,000 | 11.0 |
| 2,500,000 | 12.3 |
| 3,000,000 | 13.5 |
| 3,500,000 | 14.6 |
| 4,000,000 | 15.6 |
| 5,000,000 | 17.4 |

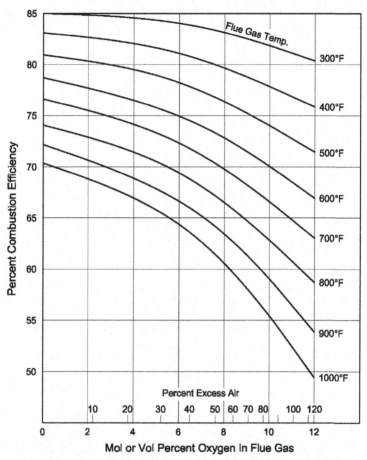

Figure 1.36 Approximate combustion efficiency of natural gas (1050 Btu/scf) in emulsion treaters.

# Heat Exchanger Configurations

**2**

▶ **OVERVIEW**

Heat exchangers used in oil and gas facilities are **configured** as follows

Bath heaters
  Direct
  Indirect
Fluid–fluid
  Shell and tube
  Double pipe
  Plate and frame
Coolers using air
  Air-cooled exchangers
  Cooling towers

The remainder of this section discusses basic concepts in **sizing and selecting** heat exchangers

▶ **SHELL-AND-TUBE EXCHANGERS**

## Tubular Exchanger Manufacturers Association (TEMA)

Defines the various types of shell and tube exchangers
Defines design and construction practices

**TEMA classes**

  Class "C"—Less stringent, used onshore, and temperatures above −20°F
  Class "R"—Offshore and cold temperature service

## Common Services
Liquid–liquid
Liquid–vapor
Vapor–vapor

## Components (Figures 2.1 and 2.2)
Shell with two nozzles
Tube sheets
Heads
Transverse baffles

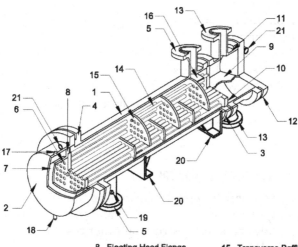

| | | |
|---|---|---|
| 1. Shell | 8. Floating Head Flange | 15. Transverse Baffles or |
| 2. Shell Cover | 9. Channel Partition |     Support Plates |
| 3. Shell Channel | 10. Stationary Tubesheet | 16. Impingement Baffle |
| 4. Shell Cover End Flange | 11. Channel | 17. Vent Connection |
| 5. Shell Nozzel | 12. Channel Cover | 18. Drain Connection |
| 6. Floating Tubesheet | 13. Channel Nozzle | 19. Test Connection |
| 7. Floating Head | 14. The Roads and Spacers | 20. Support Saddles |
| | | 21. Lifting Ring |

**Figure 2.1** Components of shell-and-tube exchangers.

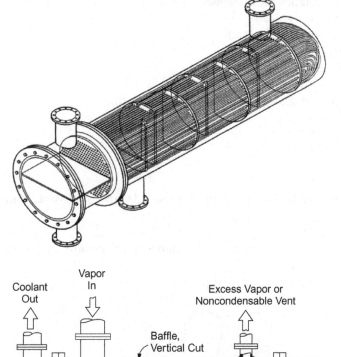

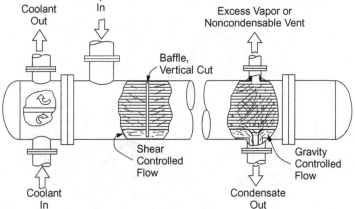

**Figure 2.2** Cutaway of two shell-and-tube heat exchangers.

## Configuration Considerations
Fluids involved
Corrosion potential
Problems of cleaning
Pressure drop
Heat transfer efficiency
**Heat exchangers selection is not routine**
Tube lengths
   20 ft
   40 ft

## Baffles
Directs the flow of both tube-side and shell-side fluids (Figures 2.3–2.6)
Pass Partition
   Forces fluid to flow through **several groups** of **parallel tubes**
   Increases number of passes

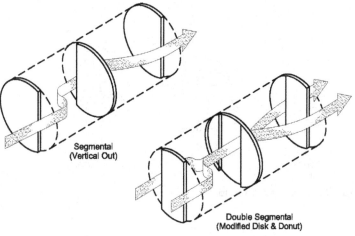

Segmental
(Vertical Out)

Double Segmental
(Modified Disk & Donut)

**Figure 2.3** Exchanger baffles.

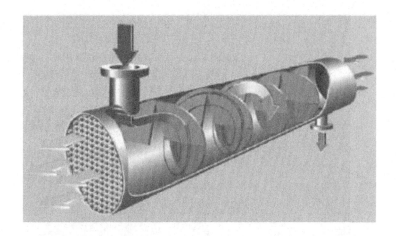

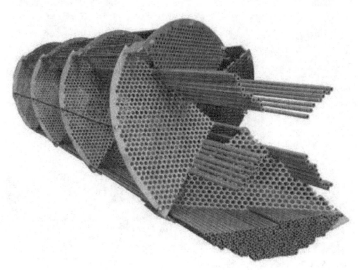

**Figure 2.4** Shell-and-tube baffles—Helical baffles.

**Figure 2.5** Shell-and-tube baffles—Helical baffles.

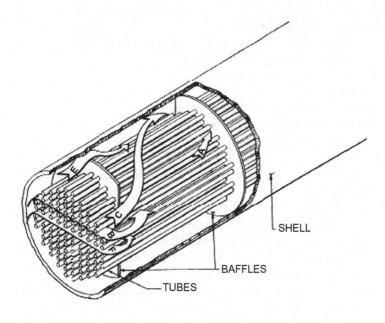

SHELL

BAFFLES

TUBES

**Streams**

**A – Tube-to-baffle leakage (partly effective)**

**B – Flow across tube bundle (most effective)**

**C – Flow around perimeter of tube bundle**

**D – Baffle-to-shell leakage**

**E – Flow through tube pass partition lanes**

**Figure 2.6** Shell-side flow paths.

Two-pass fixed exchanger
  Tube fluid passes through fraction of tubes
  Maintains high tube velocity
  Increases heat transfer
Impingement
  Placed opposite shell-side inlet nozzle
  Disperses fluid around tubes
  Prevents impingement and erosion of tubes
Transverse or support (Figures 2.7–2.10)
  **Support the tubes** that pass through holes in the baffles
  Maintains **turbulence** of shell-side fluid
  Results in **greater heat transfer**
  50% (half circle yields least $\Delta P$ but maintains support)
  Height 75% of inside diameter of shell
Up-and-down flow
Side-to-side flow (used when gas/liquid mixture flows through shell)

**U-tube heat exchanger**

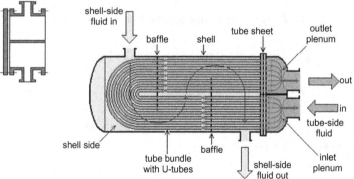

Figure 2.7 Shell-and-tube heat exchanger baffles.

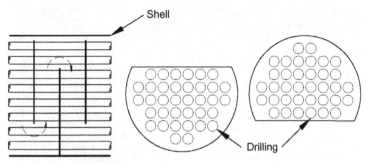

**Figure 2.8** Transverse baffle detail.

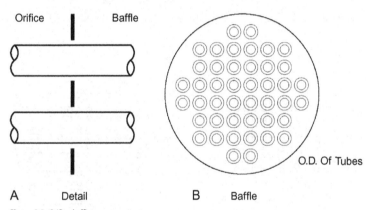

**Figure 2.9** Orifice baffle.

Longitudinal
Forces shell-side fluid to make more than one pass through the exchanger

### Use of Baffles
Main functions of a baffle are
Hold tubes in position (preventing sagging), both in production and in operation

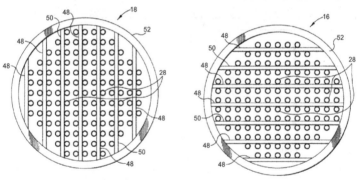

**Figure 2.10** Shell-and-tube baffles—Rod baffles.

Prevent the effects of vibration, which is increased with
   both fluid velocity and the length of the exchanger
Main functions of a baffle are:
   Direct shell-side fluid flow along tube field; this increases
      fluid velocity and the effective heat transfer coefficient
      of the exchanger

**Types of Baffles**

Implementation of baffles are decided on the basis of
  Size
  Cost
  Ability to lend support to tube bundles
  Direct flow
Often this is linked to available pressure drop and the size
  and number of passes within the exchanger
Special allowances/changes are also made for finned tubes
Different types of baffles include
  Segmental baffles (of which single segment is the most
    common)
  Rod or bar baffles (giving a uniform shell-side flow)
  Helical baffles (similar to segmental with less pressure
    drop for same size exchanger)
  Longitudinal flow baffles (used in a two-pass shell)
  Impingement baffle plates (used for protecting bundle
    when entrance velocity is high)


**Installation of Baffles**

Baffles deal with the concern of support tubes and fluid di-
  rection in heat exchangers
It is vital that they are spaced correctly at installation
The minimum baffle spacing is the greater of 50 mm or one-
  fifth the inner shell diameter
The maximum baffle spacing is dependent on material and
  size of tubes
The TEMA establishes guidelines
There are also segments with a "no tubes in window" design that affects the acceptable spacing within the design
An important design consideration is that no recirculation
  zones or dead spots form—both of which are counterpro-
  ductive to effective heat transfer

## Tubes

Not to be confused with steel pipe (extruded)

**Outside diameter is actual OD** in inches (strict tolerance)

Common tube sizes
  5/8 in.
  3/4 in.
  1 in.

Available in a variety of metals
  Steel
  Copper
  Brass
  70-30 copper nickel
  Bronze
  Aluminum
  Stainless steel

Wall thickness defined by Birmingham Wire Gauge (BWG)

## Shells

Less than 24-in. OD
  Fabricated from pipe
  Nominal diameters are used
  Wall thickness determined as in piping
Greater than 24-in. OD
  Fabricated from rolled steel plate
  Wall thickness determined from ASME Boiler and Pressure Vessel Code
  TEMA Standards specify minimum weight of shell

## Tube Pitch (Figure 2.11)

Tube holes cannot be drilled too close together as it weakens the tube sheet

"Pitch"—shortest center-to-center distance between tubes

"Clearance"—shortest distance between outside of two adjacent tubes

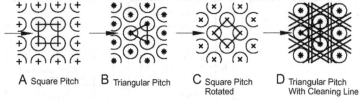

**A** Square Pitch    **B** Triangular Pitch    **C** Square Pitch Rotated    **D** Triangular Pitch With Cleaning Line

**Figure 2.11** Common tube layouts for shell-and-tube heat exchangers.

Square pitch
  Easy to clean
  Offers less $\Delta P$ when shell-side fluid flows perpendicular
    to tube axis
  Common pitches:
    ¾ in. OD × 1 in.
    1 in. OD × 1¼ in.
Triangular pitch
  Common pitches:
    ¾ in. OD × 15/16 in.
    ¾ in. OD × 1 in.
    1 in. OD × 1¼ in.

**Options**
There are many different arrangements of the shells, tubes,
  and baffles in heat exchangers
Figure 2.12 shows TEMA nomenclature for shell-and-tube
  exchangers
  Nomenclature assigns letter designations for stationary
    head, shell, and rear head
  Common combinations are shown in Figures 2.13–2.15
Three-letter "type" description
  First letter designates front end
  Second letter designates shell type
  Third letter designates back end

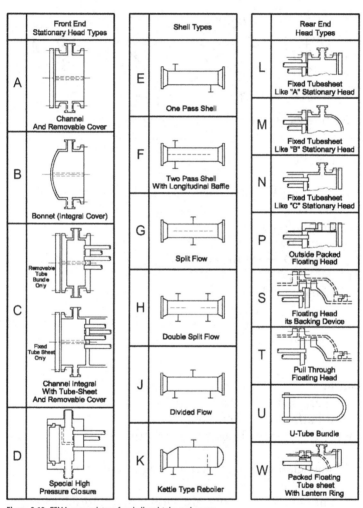

**Figure 2.12** TEMA nomenclature for shell-and-tube exchangers.

1. Stationary Head-Channel
2. Stationary Head-Bonnet
3. Stationary Head Flange-Channel Or Bonnet
4. Channel Cover
5. Stationary Head Nozzle
6. Stationary Tubesheet
7. Tubes
8. Shell
9. Shell Cover
10. Shell Flange-Stationary Head End
11. Shell Flange-Rear Head End
12. Shell Nozzle
13. Shell Cover Flange
14. Expansion Joint
15. Floating Tubesheet
16. Floating Head Cover
17. Floating Head Flange
18. Floating Head Backing Device
19. Split Shear Ring

20. Slip-On Backing Flange
21. Floating Head Cover-External
22. Floating Tubesheet Skirt
23. Packing Box
24. Packing
25. Packing Gland
26. Lantern Ring
27. Tierods & Spacers
28. Transverse Baffles Or Support Plates
29. Impingement Plate
30. Longitudinal Baffle
31. Pass Partition
32. Vent Connection
33. Drain Connection
34. Instrument Connection
35. Support Saddle
36. Lifting Lug
37. Support Bracket
38. Weir
39. Liquid Level Connection

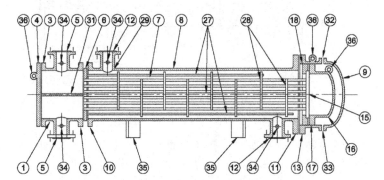

1-Pass Shell, 2-Pass Tube Exchanger

**Figure 2.13** Heat exchanger components (type AES).

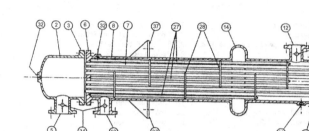

1-Pass Shell and Tube with Expansion Joint on Shell Side

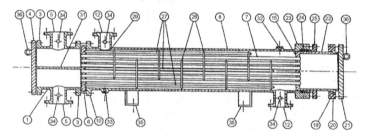

1-Pass Shell and 2-Pass Tube

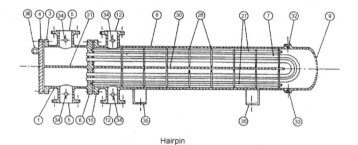

Hairpin

**Figure 2.14** Examples of tubular exchangers.

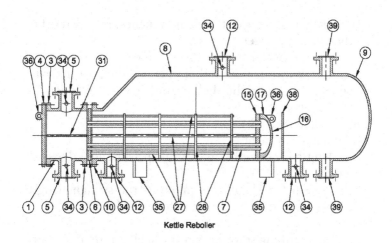

Kettle Reboiler

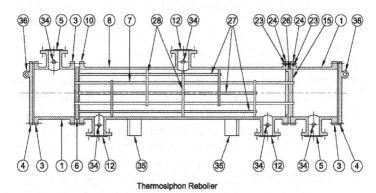

Thermosiphon Reboiler

**Figure 2.15** Common types of reboilers.

## Classification of Exchangers

Three element shorthand

First element—nominal diameter

Second element—nominal length

Third element—type

Nominal diameter

The inside diameter of the shell in inches, rounded off to the nearest integer

For kettle reboiler and chillers (that have narrow end and fat end), the nominal diameter is the port (narrow end) followed by the shell diameter, each rounded off to the nearest integer

Nominal length

Tube length in inches

Tube length for straight tubes is taken as the actual overall length

Tube length for U tubes is taken as the straight length from end of tube to bend tangent

Type

Letters describing stationary head, shell (omitted for bundles only), and rear head in that order

### Examples of Classification

**Fixed tube sheet (L)** exchanger with **removable channel and cover (A), single pass shell (E)**, 23-in. inside diameter with tubes 16 ft long is denoted Size 23-192 **Type AEL**

**Pull through floating head (T), kettle type reboiler (K)** having **stationary head** integral with **tube sheet (C)**, 23-in. port diameter and 37-in. inside shell diameter with tubes 16 ft long is denoted as Size 23/37-192 **Type CKT**

### Selection of Types

In selecting an exchanger one must know the advantages and disadvantages

Types

Fixed tube sheet

U tube

Floating head

## Selecting Heat Exchanger Components
Ensure long-term efficient heat transfer performance
  Simplify maintenance
  Prevent fouling/mechanical problems

## Picking Heat Exchanger Components
Picking shell style
  "E type"—Typical shell choice, also the least expensive;
    consider other types if there are pressure drop limitations
  "F type"—Optimum theoretical type but has problems
    with longitudinal leakage
  "J type"—Cuts flow path in half and flow rate per pass in
    half
  "X type"—Have the shortest flow path
    Used in reboilers with shell-side boiling and for services
      with very little available pressure drop
    U tubes cost the least and are the easiest to maintain
    Using a full support plate at the U-bend tangent pre-
      vents vibration

## Picking Baffles
Baffles can be one of four types (refer to Figures 2.16 and 2.17)
  Segmental
  Double segmental
  Triple segmental
  Segmental no tubes in window
For transverse (central) baffle spacing
    Start with a typical spacing of 20 to 50% shell inner di-
      ameter for sensible heat transfer
    Provide for economic gradient and good heat transfer
    Meet TEMA requirements for support and structural
      strength
For baffle cuts
  Single segmental with vertical cuts (for maintenance)
  Most efficient cut is 20 to 30% shell-side diameter

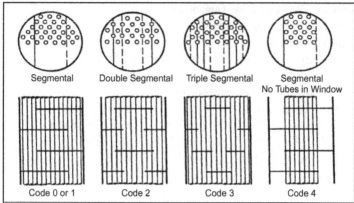

**Figure 2.16** Baffle plates.

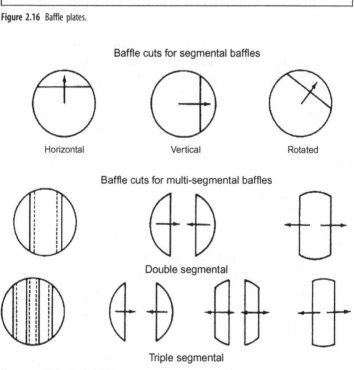

**Figure 2.17** Shell-and-tube baffles.

## Picking Components
Details are important
   Use seal bars and dummy tubes to reduce bypassing
   Use support plates to prevent tube vibration
   Include properly placed impingement rods
   Include pull holes for bundle extraction
See also Figures 2.18–2.26.

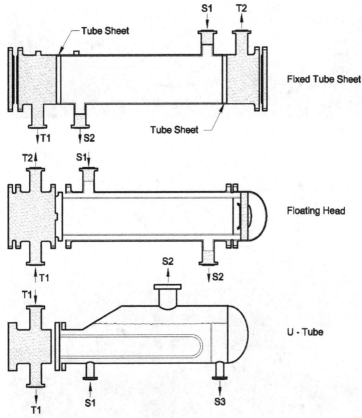

**Figure 2.18** Basic types of shell-and-tube exchangers.

Figure 2.19  TEMA type AET.

Figure 2.20  TEMA type NEN.

Figure 2.21 TEMA type BEM.

Figure 2.22 TEMA type BES (vertical).

**Figure 2.23** TEMA type BKM (double kettle).

**Figure 2.24** TEMA type CKU.

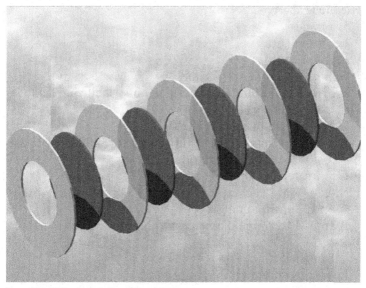

**Figure 2.25** Shell-and-tube baffles—Disk and donut.

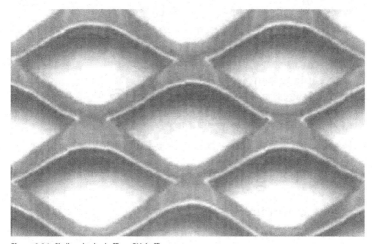

**Figure 2.26** Shell-and-tube baffles—EM baffles.

## Fixed Tube Sheet

Advantages
  Least expense
  Few gaskets
  Can replace individual tubes
  Provides maximum protection against shell-side leakage
Disadvantages
  Cannot clean or inspect shell
  Limited to temperature difference of 200°F without
    costly shell expansion joint
  Cannot replace bundle
Heads
  A and L most common
  B and M for large diameter and high pressures
Uses
  Clean shell fluid
  Low differential temperature

## U Tube (Hairpin)

Advantages
  Low cost—may be less expensive than fixed
  Can handle thermal expansion
  No internally gasketed joint needed (good for high
    pressure)
  Can replace bundle
Disadvantages
  Tubes cannot be cleaned mechanically
  Individual tubes are not replaceable
  Fewer tubes can be installed in a given shell diameter
Heads
  B and U most common
Uses
  High temperature

Clean tube fluids

Need to minimize contamination of tube fluids with higher pressure shell fluids

**Floating Head**

Advantages

  Most versatile

  High differential temperature

  Can clean both tubes and heads mechanically

  Can replace individual tubes

Disadvantages

  Most expensive

  Higher cost (at least 25% over fixed)

  Internal gasket can leak

  Difficult to plug individual tubes

Heads

  T and W least expensive (potential of shell leakage to atmosphere)

  S medium cost

  P most expensive

Uses

  Dirty, high temperature differential service

**Placement of Fluid**

Consider placing fluid through

  **Tubes** when:

    Special alloy materials are required for corrosion control and high temperatures

    Fluid is at high pressure (minimizes cost)

    Fluid contains vapors and noncondensable gases (better heat transfer)

    Fluid is scale forming (can ream out tubes)

    Toxic and lethal fluids (minimizes leakage)

**Shell** when:

> **Small pressure drops** are desired (less Δp going through shell)
>
> Fluid is **viscous** (less Δp and better heat transfer)
>
> Fluid is **nonfouling** (harder to clean shell)
>
> **Condensing or boiling service** (kettle design best)
>
> Fluid has **low flow rate** and **nonfouling** (finned tubes can be used)

### Mean Temperature Difference (MTD) Correction Factor

"F" factor is unity in MTD equation

> **Double-pipe** exchangers
>
> **Countercurrent** flow **shell-and-tube** exchangers with an **equal number** of shell and tube passes

The way fluid flows through the exchanger will affect the MTD

Correction factor function of

> Number of **tube passes**
>
> Number of **shell passes**

It is seldom that a configuration should be chosen where "F" is less than 0.8.

Figures 2.27 and 2.28 provide a means to estimate the "F" factor.

### Corrected MTD

General equation

$$MTD = F \left[ \frac{\Delta T_1 - \Delta T_2}{l_N \left( \frac{\Delta T_1}{\Delta T_2} \right)} \right] \qquad (2.1)$$

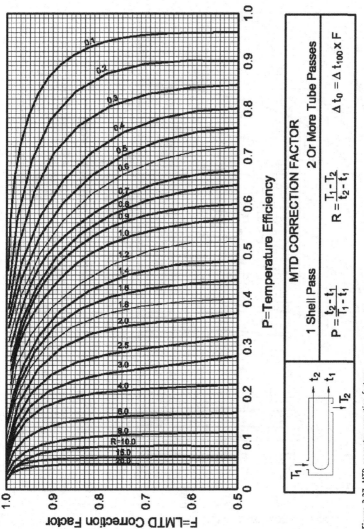

Figure 2.27  MTD geometry correction factors.

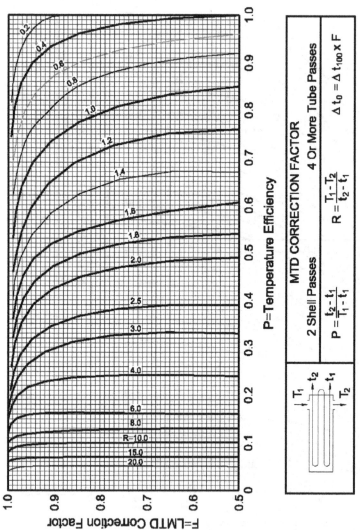

Figure 2.27—Cont'd

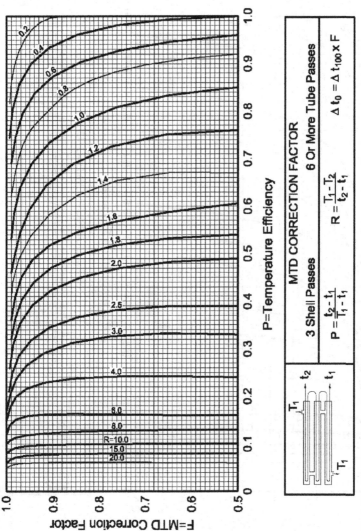

Figure 2.28 MTD geometry correction factors.

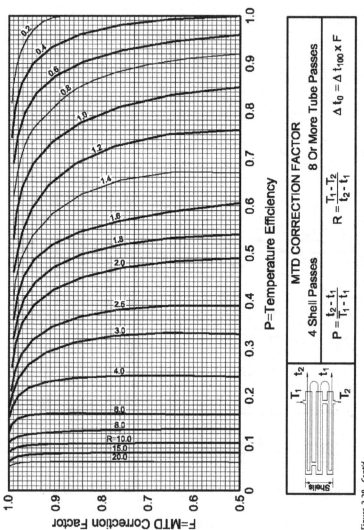

Figure 2.28—Cont'd

where

$T_1$    = hot fluid inlet temperature, °F

$T_2$    = hot fluid outlet temperature, °F

$T_3$    = cold fluid inlet temperature, °F

$T_4$    = cold fluid outlet temperature, °F

$\Delta T_1$    = larger temperature difference, °F

$\Delta T_2$    = smaller temperature difference, °F

$MTD$    = mean temperature difference, °F

$F$    = correction factor (from **Figures 2.27 and 2.28**)

### Heat Exchanger Specification Sheet

Figure 2.29 is an example of a TEMA heat exchanger specification sheet

### Sizing Procedures

Calculate **heat duty**

Determine **which fluid** will be in the **shell** and which fluid will be in the **tube**

Calculate/assume **overall heat transfer coefficient**

Select the

Number of shell passes

Number of tube passes

Calculate correction factor and corrected MTD

Choose tube diameter and tube length

Calculate number of tubes required from the following

$$N = \frac{q}{UA^1(MTD)L} \qquad (2.2)$$

| # | | | | | | | | |
|---|---|---|---|---|---|---|---|---|
| 1 | | | | | | Job No. | | |
| 2 | Customer | | | | | Reference No. | | |
| 3 | Address | | | | | Proposal No. | | |
| 4 | Plant Location | | | | | Date | | Rev. |
| 5 | Service of Unit | | | | | Item No. | | |
| 6 | Size | | Type | (Hor/Vert) | | Connected  In | Parallel | Series |
| 7 | Surf/Unit (Gross / Eff.) | | | Sq  Ft  Shells/Unit | | Surf/Shell (Gross/Eff.) | | Sq  Ft |
| 8 | PERFORMANCE OF ONE UNIT | | | | | | | |
| 9 | Fluid Allocation | | | Shell Side | | Tube Side | | |
| 10 | Fluid Name | | | | | | | |
| 11 | Fluid Quantity, Total | | Lb/Hr | | | | | |
| 12 | Vapor (In/Out) | | | | | | | |
| 13 | Liquid | | | | | | | |
| 14 | Steam | | | | | | | |
| 15 | Water | | | | | | | |
| 16 | Noncondensable | | | | | | | |
| 17 | Temperature (In/Out) | | °F | | | | | |
| 18 | Specific Gravity | | | | | | | |
| 19 | Viscosity, Liquid | | Cp | | | | | |
| 20 | Molecular Weight, Vapor | | | | | | | |
| 21 | Molecular Weight, Noncondensable | | | | | | | |
| 22 | Specific Heat | | Btu/Lb °F | | | | | |
| 23 | Thermal Conductivity | | Btu  Ft/Hr Sq Ft °F | | | | | |
| 24 | Latent Heat | | Btu/Lb @ °F | | | | | |
| 25 | Inlet Pressure | | Psig | | | | | |
| 26 | Velocity | | Ft/S | | | | | |
| 27 | Pressure Drop, Allow./Calc. | | Psi | / | | | / | |
| 28 | Fouling Resistance (Min.) | | | | | | | |
| 29 | Hear Exchanged | | | Btu/Hr: MTD (Corrected) | | | | °F |
| 30 | Transfer Rate. Service | | | Clean | | | | Btu/Hr Sq Ft  °F |
| 31 | CONSTRUCTION OF ONE SHELL | | | | | Sketch (Bundle/Nozzle Orientation) | | |
| 32 | | | Shell Side | | Tube Side | | | |
| 33 | Design/Test Pressure | Psig | / | | / | | | |
| 34 | Design Temperature | °F | | | | | | |
| 35 | No. Passes per Shell | | | | | | | |
| 36 | Corrosion Allowance | In. | | | | | | |
| 37 | Connections | In | | | | | | |
| 38 | Size & | Out | | | | | | |
| 39 | Rating | Intermediate | | | | | | |
| 40 | Tube No. | OD | In.:Thk (Min/Avg) | | In.: Length    Ft: Pitch | In.  ⬦ 30  ⬦ 60 ⬦ 90 ⬦ 45 | | |
| 41 | Tube Type | | | | Material | | | |
| 42 | Shell | ID | OD | In. | Shell Cover | | (Integ.) (Remov.) | |
| 43 | Channel or Bonnet | | | | Channel Cover | | | |
| 44 | Tubesheet-Stationary | | | | Tubesheet-Floating | | | |
| 45 | Floating Head Cover | | | | Impingement Protection | | | |
| 46 | Baffles-Cross | | Type | | % Cut (Diam/Area) | Spacing: c/c | Inlet | In. |
| 47 | Baffles-Long | | | | Seal Type | | | |
| 48 | Supports-Tube | | | U-Bend | | Type | | |
| 49 | Bypass Seal Arrangement | | | | Tube-Tubesheet Joint | | | |
| 50 | Expansion Joint | | | | Type | | | |
| 51 | ρv² - Inlet Nozzle | | | Bundle Entrance | | Bundle Exit | | |
| 52 | Gaskets-Shell Side | | | | Tube Side | | | |
| 53 | -Floating Head | | | | | | | |
| 54 | Code Requirements | | | | TEMA Class | | | |
| 55 | Weight/Shell | | | Filled with Water | | Bundle | | Lb |
| 56 | Remarks | | | | | | | |
| 57 | | | | | | | | |
| 58 | | | | | | | | |
| 59 | | | | | | | | |
| 60 | | | | | | | | |
| 61 | | | | | | | | |

**Figure 2.29** TEMA shell-and-tube exchanger specification sheet.

where

| | | |
|---|---|---|
| $N$ | = | required number of tubes |
| $q$ | = | heat duty, Btu/hr |
| $U$ | = | overall heat transfer coefficient, Btu/hr ft$^2$ °F |
| $LMTD$ | = | Corrected log mean temperature difference, °F |
| $L$ | = | Tube length, ft. |
| $A^1$ | = | Tube external surface area per foot of length, ft$^2$/ft (from Table 2.1) |

Select a shell diameter that can accommodate the number of tubes required (Figure 2.30 and Tables 2.2–2.6).

**Figure 2.30** lists the total number of tubes, not the number of tubes per pass

Multiple pass units have fewer total tubes due to partition plate

Floating head has fewer tubes than fixed head because heads and seals restrict the use of space

U tubes have the lowest number of tubes because of the space required for the tightest radius bend in the U-tube bundle

Once the number of tubes is determined, the **flow velocity of fluid inside the tubes** should be **checked**

## Example 2

Given:

| | |
|---|---|
| **Geometry** | 483¾-in. tubes, 20 ft long |
| F | 1.0 (pure counterflow) |
| $W_{hot}$ | 100,000 lbs/hr |
| $W_{cold}$ | 500,000 lbs/hr |
| $C_{p,hot}$ | 0.77 Btu/lb-°F |
| $T_{hot,in}$ | 240°F |
| $T_{hot,out}$ | 120°F |
| $T_{cold,in}$ | 90°F |

**TABLE 2.1** Characteristics of Tubing

| TUBE OD IN. | B.W.G. GAUGE | THICKNESS IN. | INTERNAL AREA IN. | FT EXTERNAL SURFACE PER FT LENGTH | FT INTERNAL SURFACE PER FT LENGTH |
|---|---|---|---|---|---|
| 1/4 | 22 | .028 | .0295 | .0655 | .0508 |
| 1/4 | 24 | .022 | .0333 | .0655 | .0539 |
| 1/4 | 26 | .018 | .0360 | .0655 | .0560 |
| 1/4 | 27 | .016 | .0373 | .0655 | .0570 |
| 3/8 | 18 | .049 | .0603 | .0982 | .0725 |
| 3/8 | 20 | .035 | .0731 | .0982 | .0798 |
| 3/8 | 22 | .028 | .0799 | .0982 | .0835 |
| 3/8 | 24 | .022 | .0860 | .0982 | .0867 |
| 1/2 | 16 | .065 | .1075 | .1309 | .0969 |
| 1/2 | 18 | .049 | .1269 | .1309 | .1052 |
| 1/2 | 20 | .035 | .1452 | .1309 | .1126 |
| 1/2 | 22 | .028 | .1548 | .1636 | .1162 |
| 5/8 | 12 | .109 | .1301 | .1636 | .1066 |
| 5/8 | 13 | .095 | .1486 | .1636 | .1139 |
| 5/8 | 14 | .083 | .1655 | .1636 | .1202 |
| 5/8 | 15 | .072 | .1817 | .1636 | .1259 |
| 5/8 | 16 | .065 | .1924 | .1636 | .1296 |
| 5/8 | 17 | .058 | .2035 | .1636 | .1333 |
| 5/8 | 18 | .049 | .2181 | .1636 | .1380 |
| 5/8 | 19 | .042 | .2298 | .1636 | .1416 |
| 5/8 | 20 | .035 | .2419 | .1963 | .1453 |

| | | | | | |
|---|---|---|---|---|---|
| 3/4 | 10 | .123 | .1825 | .1963 | .1262 |
| 3/4 | 11 | .120 | .2043 | .1963 | .1335 |
| 3/4 | 12 | .109 | .2223 | .1963 | .1393 |
| 3/4 | 13 | .095 | .2463 | .1963 | .1466 |
| 3/4 | 14 | .083 | .2679 | .1963 | .1529 |
| 3/4 | 15 | .072 | .2884 | .1963 | .1587 |
| 3/4 | 16 | .065 | .3019 | .1963 | .1623 |
| 3/4 | 17 | .058 | .3157 | .1963 | .1660 |
| 3/4 | 18 | .049 | .3339 | .1963 | .1707 |
| 3/4 | 20 | .035 | .3526 | .1963 | .1780 |
| 1 | 8 | .165 | .3632 | .2618 | .1754 |
| 1 | 10 | .134 | .4208 | .2618 | .1916 |
| 1 | 11 | .120 | .4536 | .2618 | .1990 |
| 1 | 12 | .109 | .4803 | .2618 | .2047 |
| 1 | 13 | .095 | .5153 | .2618 | .2121 |
| 1 | 14 | .083 | .5463 | .2618 | .2183 |
| 1 | 15 | .072 | .5755 | .2618 | .2241 |
| 1 | 16 | .065 | .5945 | .2618 | .2278 |
| 1 | 18 | .049 | .6390 | .2618 | .2361 |
| 1 | 20 | .035 | .6793 | .2618 | .2435 |

(Continued)

**TABLE 2.1** Characteristics of Tubing—*cont'd*

| TUBE OD IN. | B.W.G. GAUGE | THICKNESS IN. | INTERNAL AREA IN. | FT EXTERNAL SURFACE PER FT LENGTH | FT INTERNAL SURFACE PER FT LENGTH |
|---|---|---|---|---|---|
| 11/4 | 7 | .180 | .6221 | .3272 | .2330 |
| 11/4 | 8 | .165 | .6648 | .3272 | .2409 |
| 11/4 | 10 | .134 | .7574 | .3272 | .2571 |
| 11/4 | 11 | .120 | .8012 | .3272 | .2644 |
| 11/4 | 12 | .109 | .8365 | .3272 | .2702 |
| 11/4 | 13 | .095 | .8825 | .3272 | .2775 |
| 11/4 | 14 | .083 | .9229 | .3272 | .2838 |
| 11/4 | 16 | .065 | .9852 | .3272 | .2932 |
| 11/4 | 18 | .049 | 1.042 | .3272 | .3016 |
| 11/4 | 20 | .035 | 1.094 | .3272 | .3089 |
| 11/2 | 10 | .134 | 1.192 | .3927 | .3225 |
| 11/2 | 12 | .109 | 1.291 | .3927 | .3356 |
| 11/2 | 14 | .083 | 1.398 | .3927 | .3492 |
| 11/2 | 16 | .065 | 1.474 | .3927 | .3587 |
| 2 | 11 | .120 | 2.433 | .5236 | .4608 |
| 2 | 12 | .109 | 2.494 | .5236 | .4665 |
| 2 | 13 | .095 | 2.573 | .5236 | .4739 |
| 2 | 14 | .083 | 2.642 | .5236 | .4801 |

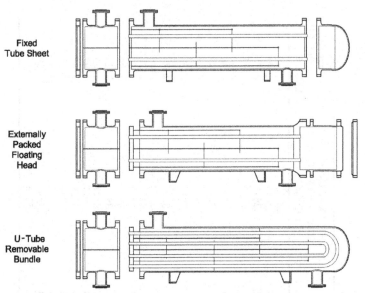

**Figure 2.30** Heat exchanger tube count.

Calculate:
  MTD
  U
Soution:
  1) Calculate area
  A = nπdl
    = 483π(0.75/12)(20)
    = 1897 ft$^2$
  2) Calculate duty
  $Q_h$ = $WC_{p,hot}\Delta T_h$
    = (100.000)(0.77)(240-120)
    = 9,240,000 Btu/hr

**TABLE 2.2** Heat Exchanger Tube Count

| TUBE OD IN. | B.W.G GAUGE | THICKNESS IN. | INTERNAL AREA IN. | FT EXTERNAL SURFACE PER FT LENGTH | FT EXTERNAL SURFACE PER FT LENGTH |
|---|---|---|---|---|---|
| 1 | 14 | .083 | .5463 | .2618 | .2183 |
| 1 | 15 | .072 | .5755 | .2618 | .2241 |
| 1 | 16 | .065 | .5945 | .2618 | .2278 |
| 1 | 18 | .049 | .6390 | .2618 | .2361 |
| 1 | 20 | .035 | .6793 | .2618 | .2435 |
| 1 1/4 | 7 | .180 | .6221 | .3272 | .2330 |
| 1 1/4 | 8 | .165 | .6648 | .3272 | .2409 |
| 1 1/4 | 10 | .134 | .7574 | .3272 | .2571 |
| 1 1/4 | 11 | .120 | .8012 | .3272 | .2644 |
| 1 1/4 | 12 | .109 | .8365 | .3272 | .2702 |
| 1 1/4 | 13 | .095 | .8825 | .3272 | .2775 |
| 1 1/4 | 14 | .083 | .9229 | .3272 | .2838 |
| 1 1/4 | 16 | .065 | .9852 | .3272 | .2932 |
| 1 1/4 | 18 | .049 | 1.042 | .3272 | .3016 |
| 1 1/4 | 20 | .035 | 1.094 | .3272 | .3089 |
| 1 1/2 | 10 | .134 | 1.192 | .3927 | .3225 |
| 1 1/2 | 12 | .109 | 1.291 | .3927 | .3356 |
| 1 1/2 | 14 | .083 | 1.398 | .3927 | .3492 |
| 1 1/2 | 16 | .065 | 1.474 | .3927 | .3587 |
| 2 | 11 | .120 | 2.433 | .5236 | .4608 |
| 2 | 12 | .109 | 2.494 | .5236 | .4665 |
| 2 | 13 | .095 | 2.573 | .5236 | .4739 |
| 2 | 14 | .083 | 2.642 | .5236 | .4801 |

**TABLE 2.3** Heat Exchanger Tube Count

| | 3/4" OD TUBES ON 1" □ PITCH | | | | | | | |
|---|---|---|---|---|---|---|---|---|
| | FIXED TUBE SHEET | | | OUTSIDE PACKED FLOATING HEAD | | | U-TUBE | |
| | NO. OF PASSES | | | NO. OF PASSES | | | NO. OF PASSES | |
| SHELL ID (INCHES) | 1 | 2 | 4 | 1 | 2 | 4 | 2 | 4 |
| 5.047 | 12 | 12 | 12 | 12 | 6 | 4 | 3 | 2 |
| 6.065 | 21 | 16 | 16 | 16 | 16 | 12 | 4 | 4 |
| 7.981 | 37 | 34 | 32 | 32 | 28 | 24 | 12 | 10 |
| 10.02 | 61 | 60 | 52 | 52 | 52 | 52 | 22 | 20 |
| 12.00 | 97 | 88 | 88 | 81 | 78 | 76 | 34 | 34 |
| 13.25 | 112 | 112 | 112 | 97 | 94 | 88 | 45 | 44 |
| 15.25 | 156 | 148 | 148 | 140 | 132 | 124 | 64 | 60 |
| 17.25 | 208 | 196 | 188 | 188 | 178 | 172 | 88 | 84 |
| 19.25 | 250 | 249 | 244 | 241 | 224 | 216 | 112 | 108 |
| 21.25 | 316 | 307 | 296 | 296 | 280 | 276 | 138 | 134 |
| 23.25 | 378 | 370 | 370 | 356 | 344 | 332 | 170 | 166 |
| 25.00 | 442 | 432 | 428 | 414 | 406 | 392 | 200 | 194 |
| 27.00 | 518 | 509 | 496 | 482 | 476 | 468 | 236 | 230 |
| 29.00 | 602 | 596 | 580 | 570 | 562 | 548 | 277 | 272 |
| 31.00 | 686 | 676 | 676 | 658 | 640 | 640 | 320 | 312 |
| 33.00 | 782 | 768 | 768 | 742 | 732 | 732 | 362 | 360 |
| 35.00 | 896 | 868 | 868 | 846 | 831 | 820 | 418 | 406 |
| 37.00 | 1004 | 978 | 964 | 952 | 931 | 928 | 470 | 462 |

(Continued)

TABLE 2.3 Heat Exchanger Tube Count—cont'd

## 3/4" OD TUBES ON 1" □ PITCH

| SHELL ID (INCHES) | FIXED TUBE SHEET | | | OUTSIDE PACKED FLOATING HEAD | | | U-TUBE | |
|---|---|---|---|---|---|---|---|---|
| | NO. OF PASSES | | | NO. OF PASSES | | | NO. OF PASSES | |
| | 1 | 2 | 4 | 1 | 2 | 4 | 2 | 4 |
| 39.00 | 1102 | 1096 | 1076 | 1062 | 1045 | 1026 | 524 | 520 |
| 42.00 | 1283 | 1289 | 1270 | 1232 | 1222 | 1218 | 611 | 602 |
| 45.00 | 1484 | 1472 | 1456 | 1424 | 1415 | 1386 | 710 | 700 |
| 48.00 | 1701 | 1691 | 1670 | 1636 | 1634 | 1602 | 812 | 802 |
| 51.00 | 1928 | 1904 | 1888 | 1845 | 1832 | 1818 | 926 | 910 |
| 54.00 | 2154 | 2138 | 2106 | 2080 | 2066 | 2044 | 1042 | 1032 |
| 60.00 | 2683 | 2650 | 2636 | 2582 | 2566 | 2556 | 1298 | 1282 |

## 3/4" OD TUBES ON 1" ◆ PITCH

| SHELL ID (INCHES) | FIXED TUBE SHEET | | | OUTSIDE PACKED FLOATING HEAD | | | U-TUBE | |
|---|---|---|---|---|---|---|---|---|
| | NO. OF PASSES | | | NO. OF PASSES | | | NO. OF PASSES | |
| | 1 | 2 | 4 | 1 | 2 | 4 | 2 | 4 |
| 5.047 | 12 | 10 | 8 | 12 | 10 | 8 | 2 | 2 |
| 6.065 | 21 | 18 | 16 | 16 | 12 | 8 | 5 | 4 |
| 7.981 | 37 | 32 | 28 | 32 | 28 | 24 | 12 | 10 |

| | | | | | | | | |
|---|---|---|---|---|---|---|---|---|
| 10.02 | 61 | 54 | 48 | 52 | 46 | 40 | 21 | 18 |
| 12.00 | 97 | 90 | 84 | 81 | 74 | 68 | 33 | 32 |
| 13.25 | 113 | 108 | 104 | 97 | 92 | 84 | 43 | 40 |
| 15.25 | 156 | 146 | 136 | 140 | 134 | 128 | 62 | 58 |
| 17.25 | 208 | 196 | 184 | 188 | 178 | 168 | 87 | 82 |
| 19.25 | 256 | 244 | 236 | 241 | 228 | 216 | 109 | 104 |
| 21.25 | 314 | 299 | 294 | 300 | 286 | 272 | 136 | 130 |
| 23.25 | 379 | 363 | 352 | 359 | 343 | 328 | 267 | 160 |
| 25.00 | 448 | 432 | 416 | 421 | 404 | 392 | 195 | 190 |
| 27.00 | 522 | 504 | 486 | 489 | 472 | 456 | 234 | 226 |
| 29.00 | 603 | 583 | 568 | 575 | 556 | 540 | 275 | 266 |
| 31.00 | 688 | 667 | 654 | 660 | 639 | 624 | 313 | 304 |
| 33.00 | 788 | 770 | 756 | 749 | 728 | 708 | 360 | 350 |
| 35.00 | 897 | 873 | 850 | 846 | 826 | 804 | 409 | 398 |
| 37.00 | 1009 | 983 | 958 | 952 | 928 | 908 | 464 | 452 |
| 39.00 | 1118 | 1092 | 1066 | 1068 | 1041 | 1016 | 518 | 508 |
| 42.00 | 1298 | 1269 | 1250 | 1238 | 1216 | 1196 | 610 | 596 |
| 45.00 | 1500 | 1470 | 1440 | 1424 | 1407 | 1378 | 706 | 692 |
| 48.00 | 1714 | 1681 | 1650 | 1644 | 1611 | 1580 | 804 | 788 |
| 51.00 | 1939 | 1903 | 1868 | 1864 | 1837 | 1804 | 917 | 902 |
| 54.00 | 2173 | 2135 | 2098 | 2098 | 2062 | 2026 | 1036 | 1018 |
| 60.00 | 2692 | 2651 | 2612 | 2600 | 2560 | 2520 | 1292 | 1272 |

**TABLE 2.4** Heat Exchanger Tube Count

1" OD TUBES ON 1–1/4" △ PITCH

| SHELL ID (INCHES) | FIXED TUBE SHEET NO. OF PASSES | | | OUTSIDE PACKED FLOATING HEAD NO. OF PASSES | | | U-TUBE NO. OF PASSES | |
|---|---|---|---|---|---|---|---|---|
| | 1 | 2 | 4 | 1 | 2 | 4 | 2 | 4 |
| 5.047 | 8 | 6 | 4 | 7 | 4 | 4 | 0 | 0 |
| 6.065 | 14 | 14 | 8 | 10 | 10 | 4 | 2 | 2 |
| 7.981 | 26 | 26 | 16 | 22 | 18 | 16 | 7 | 4 |
| 10.02 | 42 | 40 | 36 | 38 | 36 | 28 | 13 | 12 |
| 12.00 | 64 | 61 | 56 | 56 | 52 | 48 | 22 | 18 |
| 13.25 | 75 | 76 | 72 | 73 | 72 | 60 | 28 | 26 |
| 15.25 | 110 | 106 | 100 | 100 | 98 | 88 | 43 | 38 |
| 17.25 | 147 | 138 | 128 | 130 | 126 | 116 | 57 | 52 |
| 19.25 | 184 | 175 | 168 | 170 | 162 | 148 | 76 | 68 |
| 21.25 | 227 | 220 | 212 | 212 | 201 | 188 | 96 | 88 |
| 23.25 | 280 | 265 | 252 | 258 | 250 | 232 | 116 | 110 |
| 25.00 | 316 | 313 | 294 | 296 | 294 | 276 | 135 | 128 |
| 27.00 | 371 | 370 | 358 | 355 | 346 | 328 | 161 | 152 |
| 29.00 | 434 | 424 | 408 | 416 | 408 | 392 | 189 | 182 |
| 31.00 | 503 | 489 | 468 | 475 | 466 | 446 | 222 | 212 |
| 33.00 | 576 | 558 | 534 | 544 | 529 | 510 | 254 | 246 |
| 35.00 | 643 | 634 | 604 | 619 | 604 | 582 | 289 | 280 |
| 37.00 | 738 | 709 | 468 | 696 | 679 | 660 | 330 | 316 |

1" OD TUBES ON 1-1/4" □ PITCH

| SHELL ID (INCHES) | FIXED TUBE SHEET NO. OF PASSES | | | OUTSIDE PACKED FLOATING HEAD NO. OF PASSES | | | U-TUBE NO. OF PASSES | |
|---|---|---|---|---|---|---|---|---|
| | 1 | 2 | 4 | 1 | 2 | 4 | 2 | 4 |
| 39.00 | 804 | 787 | 772 | 768 | 753 | 730 | 370 | 356 |
| 42.00 | 946 | 928 | 898 | 908 | 891 | 860 | 436 | 418 |
| 45.00 | 1087 | 1069 | 1042 | 1041 | 1017 | 990 | 505 | 490 |
| 48.00 | 1240 | 1230 | 1198 | 1189 | 1182 | 1152 | 578 | 562 |
| 51.00 | 1397 | 1389 | 1354 | 1348 | 1337 | 1300 | 661 | 642 |
| 54.00 | 1592 | 1561 | 1530 | 1531 | 1503 | 1462 | 748 | 726 |
| 60.00 | 1969 | 1945 | 1904 | 1906 | 1979 | 1842 | 933 | 914 |
| 5.047 | 9 | 6 | 4 | 5 | 4 | 4 | 0 | 0 |
| 6.065 | 12 | 12 | 12 | 12 | 6 | 4 | 2 | 2 |
| 7.981 | 22 | 20 | 16 | 21 | 16 | 16 | 6 | 4 |
| 10.02 | 38 | 38 | 32 | 32 | 32 | 32 | 12 | 10 |
| 12.00 | 56 | 56 | 52 | 52 | 52 | 44 | 19 | 18 |
| 13.25 | 69 | 66 | 66 | 61 | 60 | 52 | 25 | 24 |
| 15.25 | 97 | 90 | 88 | 89 | 84 | 80 | 36 | 34 |
| 17.25 | 129 | 124 | 120 | 113 | 112 | 112 | 49 | 48 |
| 19.25 | 164 | 158 | 148 | 148 | 144 | 140 | 64 | 62 |
| 21.25 | 202 | 191 | 184 | 178 | 178 | 172 | 83 | 78 |
| 23.25 | 234 | 234 | 222 | 216 | 216 | 208 | 100 | 98 |
| 25.00 | 272 | 267 | 264 | 258 | 256 | 256 | 120 | 116 |

(Continued)

**TABLE 2.4** Heat Exchanger Tube Count—*cont'd*

1" OD TUBES ON 1-1/4" □ PITCH

| SHELL ID (INCHES) | FIXED TUBE SHEET NO. OF PASSES | | | | OUTSIDE PACKED FLOATING HEAD NO. OF PASSES | | | | U-TUBE NO. OF PASSES | |
|---|---|---|---|---|---|---|---|---|---|---|
| | 1 | 2 | 4 | | 1 | 2 | 4 | | 2 | 4 |
| 27.00 | 328 | 317 | 310 | | 302 | 300 | 296 | | 142 | 138 |
| 29.00 | 378 | 370 | 370 | | 356 | 353 | 338 | | 166 | 166 |
| 31.00 | 434 | 428 | 428 | | 414 | 406 | 392 | | 145 | 192 |
| 33.00 | 496 | 484 | 484 | | 476 | 460 | 260 | | 221 | 218 |
| 35.00 | 554 | 553 | 532 | | 542 | 530 | 518 | | 254 | 248 |
| 37.00 | 628 | 612 | 608 | | 602 | 596 | 580 | | 287 | 280 |
| 39.00 | 708 | 682 | 682 | | 676 | 649 | 648 | | 322 | 314 |
| 42.00 | 811 | 811 | 804 | | 782 | 780 | 768 | | 379 | 374 |
| 45.00 | 940 | 931 | 918 | | 904 | 894 | 874 | | 436 | 434 |
| 48.00 | 1076 | 1061 | 1040 | | 1034 | 1027 | 1012 | | 501 | 494 |
| 51.00 | 1218 | 1202 | 1192 | | 1178 | 1155 | 1150 | | 573 | 570 |
| 54.00 | 1370 | 1354 | 1350 | | 1322 | 1307 | 1284 | | 650 | 644 |
| 60.00 | 1701 | 1699 | 1684 | | 1654 | 1640 | 1632 | | 810 | 802 |

**TABLE 2.5** Heat Exchanger Tube Count

1" OD TUBES ON 1-1/4" △ PITCH

| SHELL ID (INCHES) | FIXED TUBE SHEET NO. OF PASSES | | | OUTSIDE PACKED FLOATING HEAD NO. OF PASSES | | | U-TUBE NO. OF PASSES | |
|---|---|---|---|---|---|---|---|---|
| | 1 | 2 | 4 | 1 | 2 | 4 | 2 | 4 |
| 5.047 | 8 | 6 | 4 | 5 | 4 | 4 | 0 | 0 |
| 6.065 | 12 | 10 | 8 | 12 | 10 | 8 | 2 | 2 |
| 7.981 | 24 | 20 | 16 | 21 | 18 | 16 | 5 | 4 |
| 10.02 | 37 | 32 | 28 | 32 | 32 | 28 | 12 | 10 |
| 12.00 | 57 | 53 | 48 | 52 | 46 | 40 | 18 | 16 |
| 13.25 | 70 | 70 | 64 | 61 | 58 | 56 | 25 | 22 |
| 15.25 | 97 | 90 | 84 | 89 | 82 | 76 | 35 | 32 |
| 17.25 | 129 | 120 | 112 | 113 | 112 | 104 | 48 | 44 |
| 19.25 | 162 | 152 | 142 | 148 | 138 | 128 | 62 | 60 |
| 21.25 | 205 | 193 | 184 | 180 | 174 | 168 | 78 | 76 |
| 23.25 | 238 | 228 | 220 | 221 | 210 | 200 | 100 | 94 |
| 25.00 | 275 | 264 | 256 | 261 | 248 | 236 | 116 | 110 |
| 27.00 | 330 | 315 | 300 | 308 | 296 | 286 | 141 | 134 |
| 29.00 | 379 | 363 | 360 | 359 | 345 | 336 | 165 | 160 |
| 31.00 | 435 | 422 | 410 | 418 | 401 | 388 | 191 | 184 |
| 33.00 | 495 | 478 | 472 | 477 | 460 | 448 | 220 | 212 |
| 35.00 | 556 | 552 | 538 | 540 | 526 | 508 | 249 | 242 |
| 37.00 | 632 | 613 | 598 | 608 | 588 | 568 | 281 | 274 |

(Continued)

**TABLE 2.5** Heat Exchanger Tube Count—cont'd

### 1" OD TUBES ON 1-1/4" △ PITCH

| SHELL ID (INCHES) | FIXED TUBE SHEET | | | OUTSIDE PACKED FLOATING HEAD | | | U-TUBE | |
|---|---|---|---|---|---|---|---|---|
| | NO. OF PASSES | | | NO. OF PASSES | | | NO. OF PASSES | |
| | 1 | 2 | 4 | 1 | 2 | 4 | 2 | 4 |
| 39.00 | 705 | 685 | 672 | 674 | 654 | 640 | 315 | 310 |
| 42.00 | 822 | 799 | 786 | 788 | 765 | 756 | 372 | 364 |
| 45.00 | 946 | 922 | 912 | 910 | 885 | 866 | 436 | 426 |
| 48.00 | 1079 | 1061 | 1052 | 1037 | 1018 | 1000 | 501 | 490 |
| 51.00 | 1220 | 1199 | 1176 | 1181 | 1160 | 1142 | 569 | 558 |
| 54.00 | 1389 | 1359 | 1330 | 1337 | 1307 | 1292 | 646 | 632 |
| 60.00 | 1714 | 1691 | 1664 | 1658 | 1626 | 1594 | 802 | 788 |

### 1-1/4" OD TUBES ON 1-9/16" △ PITCH

| SHELL ID (INCHES) | FIXED TUBE SHEET | | | OUTSIDE PACKED FLOATING HEAD | | | U-TUBE | |
|---|---|---|---|---|---|---|---|---|
| | NO. OF PASSES | | | NO. OF PASSES | | | NO. OF PASSES | |
| | 1 | 2 | 4 | 1 | 2 | 4 | 2 | 4 |
| 5.047 | 7 | 4 | 4 | 0 | 0 | 0 | 0 | 0 |
| 6.065 | 8 | 6 | 4 | 7 | 6 | 4 | 0 | 0 |
| 7.981 | 19 | 14 | 12 | 14 | 14 | 8 | 3 | 2 |

|        |     |     |      |      |      |      |      |      |
|--------|-----|-----|------|------|------|------|------|------|
| 10.02  | 6   | 7   | 16   | 20   | 22   | 20   | 26   | 29   |
| 12.00  | 10  | 11  | 28   | 36   | 37   | 34   | 38   | 42   |
| 13.25  | 14  | 16  | 36   | 44   | 44   | 44   | 48   | 52   |
| 15.25  | 22  | 24  | 48   | 62   | 64   | 60   | 68   | 69   |
| 17.25  | 30  | 32  | 72   | 78   | 85   | 78   | 84   | 92   |
| 19.25  | 40  | 43  | 96   | 102  | 109  | 104  | 110  | 121  |
| 21.25  | 52  | 57  | 116  | 130  | 130  | 128  | 138  | 147  |
| 23.25  | 66  | 69  | 144  | 152  | 163  | 156  | 165  | 174  |
| 25.00  | 76  | 81  | 172  | 184  | 184  | 184  | 196  | 196  |
| 27.00  | 92  | 98  | 208  | 216  | 221  | 224  | 226  | 237  |
| 29.00  | 110 | 116 | 242  | 252  | 262  | 256  | 269  | 280  |
| 31.00  | 128 | 134 | 280  | 302  | 302  | 294  | 313  | 313  |
| 33.00  | 148 | 155 | 318  | 332  | 345  | 332  | 346  | 357  |
| 35.00  | 172 | 178 | 364  | 383  | 392  | 386  | 401  | 416  |
| 37.00  | 194 | 202 | 412  | 429  | 442  | 432  | 453  | 461  |
| 39.00  | 220 | 226 | 460  | 479  | 493  | 478  | 493  | 511  |
| 42.00  | 260 | 267 | 544  | 557  | 576  | 570  | 579  | 596  |
| 45.00  | 306 | 313 | 628  | 640  | 657  | 662  | 673  | 687  |
| 48.00  | 350 | 360 | 728  | 745  | 756  | 758  | 782  | 790  |
| 51.00  | 400 | 411 | 832  | 839  | 859  | 860  | 871  | 896  |
| 54.00  | 454 | 465 | 940  | 959  | 964  | 968  | 994  | 1008 |
| 60.00  | 570 | 580 | 1170 | 1195 | 1199 | 1210 | 1243 | 1243 |

**TABLE 2.6** Heat Exchanger Tube Count

1-1/4" OD TUBES ON 1-9/16" □ PITCH

| SHELL ID (INCHES) | FIXED TUBE SHEET | | | OUTSIDE PACKED FLOATING HEAD | | | U-TUBE | |
|---|---|---|---|---|---|---|---|---|
| | NO. OF PASSES | | | NO. OF PASSES | | | NO. OF PASSES | |
| | 1 | 2 | 4 | 1 | 2 | 4 | 2 | 4 |
| 5.047 | 4 | 4 | 4 | 0 | 0 | 0 | 0 | 0 |
| 6.065 | 6 | 6 | 4 | 6 | 6 | 4 | 0 | 0 |
| 7.981 | 12 | 12 | 12 | 12 | 12 | 12 | 3 | 2 |
| 10.02 | 24 | 22 | 16 | 21 | 16 | 16 | 6 | 4 |
| 12.00 | 37 | 34 | 32 | 32 | 32 | 32 | 10 | 10 |
| 13.25 | 45 | 42 | 42 | 38 | 38 | 32 | 14 | 14 |
| 15.25 | 61 | 60 | 52 | 52 | 52 | 52 | 21 | 18 |
| 17.25 | 80 | 76 | 76 | 70 | 70 | 68 | 28 | 26 |
| 19.25 | 97 | 95 | 88 | 89 | 88 | 88 | 37 | 34 |
| 21.25 | 124 | 124 | 120 | 112 | 112 | 112 | 49 | 48 |
| 23.25 | 145 | 145 | 144 | 138 | 138 | 130 | 62 | 60 |
| 25.00 | 172 | 168 | 164 | 164 | 164 | 156 | 70 | 68 |
| 27.00 | 210 | 202 | 202 | 193 | 184 | 184 | 88 | 88 |
| 29.00 | 241 | 234 | 230 | 224 | 224 | 216 | 100 | 98 |
| 31.00 | 272 | 268 | 268 | 258 | 256 | 256 | 116 | 116 |
| 33.00 | 310 | 306 | 302 | 296 | 296 | 282 | 136 | 134 |
| 35.00 | 356 | 353 | 338 | 336 | 332 | 332 | 156 | 148 |
| 37.00 | 396 | 387 | 384 | 378 | 370 | 370 | 174 | 174 |

### 1-1/4" OD TUBES ON 1-9/16" ♦ PITCH

| SHELL ID (INCHES) | FIXED TUBE SHEET | | | OUTSIDE PACKED FLOATING HEAD | | | U-TUBE | |
|---|---|---|---|---|---|---|---|---|
| | NO. OF PASSES | | | NO. OF PASSES | | | NO. OF PASSES | |
| | 1 | 2 | 4 | 1 | 2 | 4 | 2 | 4 |
| 5.047 | 5 | 4 | 4 | 0 | 0 | 0 | 0 | 0 |
| 6.065 | 6 | 6 | 4 | 5 | 4 | 4 | 0 | 0 |
| 7.981 | 12 | 10 | 8 | 12 | 10 | 8 | 2 | 2 |
| 10.02 | 24 | 20 | 16 | 21 | 18 | 16 | 6 | 6 |
| 12.00 | 37 | 32 | 28 | 32 | 28 | 28 | 10 | 10 |
| 13.25 | 45 | 40 | 40 | 37 | 34 | 32 | 13 | 12 |
| 15.25 | 60 | 56 | 56 | 52 | 52 | 48 | 20 | 18 |
| 17.25 | 79 | 76 | 76 | 70 | 70 | 64 | 28 | 26 |
| 19.25 | 97 | 94 | 94 | 90 | 90 | 84 | 37 | 34 |
| 39.00 | 442 | 438 | 434 | 428 | 426 | 414 | 198 | 196 |
| 42.00 | 518 | 518 | 502 | 492 | 492 | 484 | 236 | 228 |
| 45.00 | 602 | 602 | 588 | 570 | 566 | 556 | 276 | 268 |
| 48.00 | 682 | 681 | 676 | 658 | 648 | 648 | 314 | 310 |
| 51.00 | 770 | 760 | 756 | 742 | 729 | 722 | 356 | 354 |
| 54.00 | 862 | 860 | 856 | 838 | 823 | 810 | 404 | 402 |
| 60.00 | 1084 | 1070 | 1054 | 1042 | 1034 | 1026 | 506 | 496 |

(Continued)

**TABLE 2.6** Heat Exchanger Tube Count—*cont'd*

1-1/4″ OD TUBES ON 1-9/16″ ♦ PITCH

| SHELL ID (INCHES) | FIXED TUBE SHEET | | | OUTSIDE PACKED FLOATING HEAD | | | U-TUBE | |
|---|---|---|---|---|---|---|---|---|
| | NO. OF PASSES | | | NO. OF PASSES | | | NO. OF PASSES | |
| | 1 | 2 | 4 | 1 | 2 | 4 | 2 | 4 |
| 21.25 | 124 | 116 | 112 | 112 | 108 | 104 | 48 | 44 |
| 23.25 | 148 | 142 | 136 | 140 | 138 | 128 | 60 | 56 |
| 25.00 | 174 | 166 | 160 | 162 | 162 | 156 | 71 | 68 |
| 27.00 | 209 | 202 | 192 | 191 | 188 | 184 | 85 | 82 |
| 29.00 | 238 | 232 | 232 | 221 | 215 | 208 | 100 | 96 |
| 31.00 | 275 | 264 | 264 | 281 | 249 | 244 | 114 | 110 |
| 33.00 | 314 | 307 | 300 | 300 | 286 | 280 | 134 | 128 |
| 35.00 | 359 | 345 | 334 | 341 | 330 | 320 | 153 | 148 |
| 37.00 | 401 | 387 | 380 | 384 | 372 | 360 | 173 | 168 |
| 39.00 | 442 | 427 | 424 | 428 | 412 | 404 | 195 | 190 |
| 42.00 | 572 | 506 | 500 | 497 | 484 | 472 | 228 | 224 |
| 45.00 | 603 | 583 | 572 | 575 | 562 | 552 | 271 | 264 |
| 48.00 | 682 | 669 | 660 | 660 | 648 | 640 | 309 | 302 |
| 51.00 | 777 | 762 | 756 | 743 | 728 | 716 | 354 | 346 |
| 54.00 | 875 | 857 | 850 | 843 | 822 | 812 | 401 | 392 |
| 60.00 | 1088 | 1080 | 1058 | 1049 | 1029 | 1016 | 505 | 492 |

3) Calculate cold outlet temperature

$Q_C = W_C C_{p,cold} \Delta T_C$

Rearranging and solving for $\Delta T_C$

$$\Delta T_C = Q_C/(W_C C_{p,\ cold})$$
$$T_{cold,out} = T_{cold,in} + Q_C/(C_{p,cold})$$
$$= 90 + 9{,}240{,}000/(50{,}000)(1.0)$$
$$= 108.5°F$$

4) Calculate LMTD

$$MTM = [(T_{hot,in} - T_{cold,out}) - (T_{hot,out} - T_{cold,in})]/\ln$$
$$[(T_{hot,in} - T_{cold,out})/(T_{hot,out} - T_{cold,in})]$$
$$= [(240 - 108.5) - (120 - 90)]/\ln[240 - 108.5/$$
$$(120 - 90)]$$
$$= 68.7°F$$

5) Calculate U

$Q = UAF(LMTD)$

Rearranging and solving for U

$$U = Q/(AF(LMTD)$$
$$= 9{,}240{,}000/[(1897)(1.0)(68.7)]$$
$$= 70.9 \text{ Btu.hr ft}^2 °F$$

▶ **DOUBLE-PIPE EXCHANGERS**

**Overview**

Consists of a pipe or tube inside a pipe shell (Figure 2.31)

Developed to fit applications that are too small to use
  TEMA requirements

Tubes are often finned to yield additional surface area

Advantages

  Inexpensive and readily available

  Can handle thermal expansion in U-bend construction

Disadvantages

  Limited to bare tube surface area of less than 500 ft$^2$
    (maximum 1000 ft$^2$)

  Cannot remove tube

  Cannot clean tube and shell mechanically

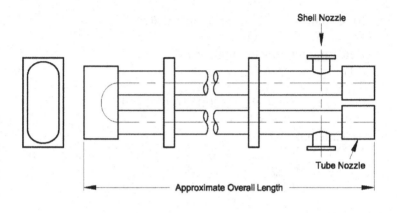

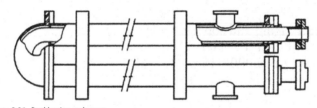

Figure 2.31 Double-pipe exchanger.

Uses
   Clean fluids
   Low heat transfer area required
   High temperature high pressure (500 psig)

## Two Shells Joined at One End Through a "Return Bonnet" (Figure 2.32)

Shell-side fluid flows in **series** in each of two shells
Results in **more compact** exchanger

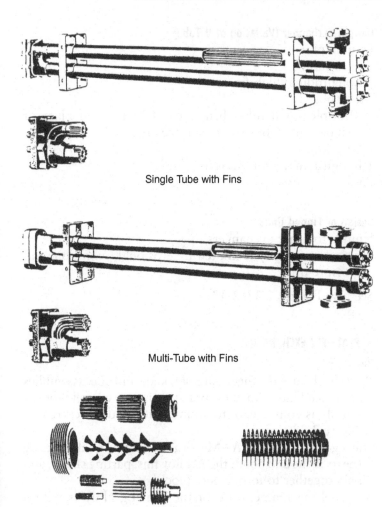

Single Tube with Fins

Multi-Tube with Fins

**Figure 2.32** Details of finned tubes and double-pipe exchangers.

## Hairpin Exchanger (Variation of U Tube)

Double pipe
  Bare tube
  Longitudinal high fin
Multitube
  Multiple small tubes bent into "U" shape in place of
    single tube (allow more surface area)
  Bare tube
  Longitudinal high fin or low fin
Banks

## Design of Finned Units

Similar to other exchangers
Maximum velocity is limited by erosional, vibration, and
  pressure drop

See also Figures 2.33–2.37.

## ▶ PLATE-FIN EXCHANGERS

### Overview

Plate-fin heat exchangers are stacked, and the assemblies
  are placed into a vacuum-brazing furnace wherein the assembly is converted into an integral, solid structure (see
  Figure 2.38)
During the process, an Al–Mg–Si alloy coating on the parting
  sheets melts and bonds the Al alloy fins, parting sheets, and
  bars together to form a core block
The melting temperature of the brazing alloy is within
  $50°F$ ($28°C$) of the melting temperature of the base
  Al alloys
The process is a very precise and highly protected process
Braised aluminum plate fins are used in low-temperature
  (cryogenic) gas processing services

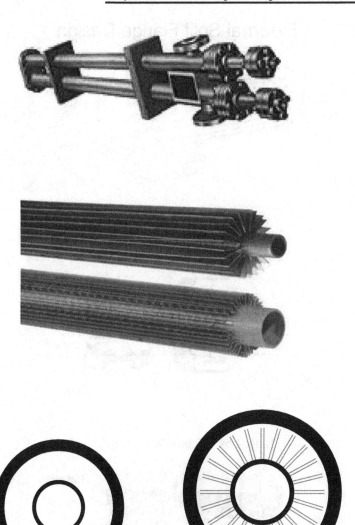

**Figure 2.33** Hairpin double-pipe exchanger; top—bare tube; bottom—longitudinal high fin.

# External Split Flange Design

Patented

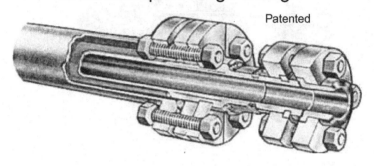

# Internal Split Ring Design

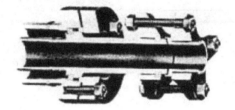

Taper-Lok ® Tubeside Closure

Patented

**Figure 2.34** Hairpin double-pipe closures.

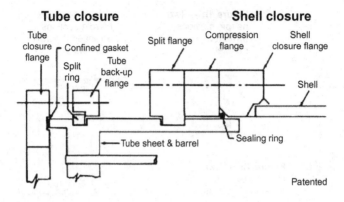

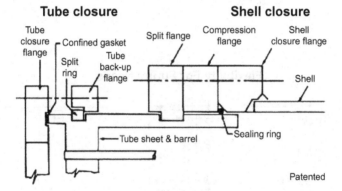

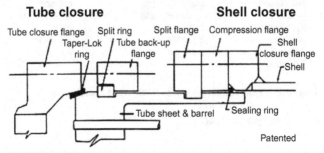

**Figure 2.35** Hairpin multitube closures.

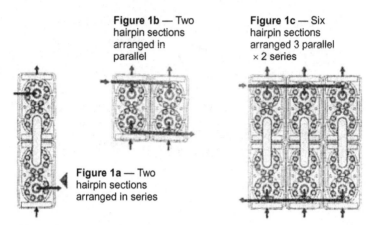

**Figure 1b** — Two hairpin sections arranged in parallel

**Figure 1c** — Six hairpin sections arranged 3 parallel × 2 series

**Figure 1a** — Two hairpin sections arranged in series

**Figure 1** — Typical arrangement of sections to meet specific duties

## Flexibility For Future Needs

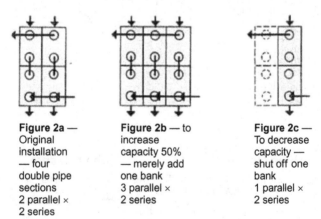

**Figure 2a** — Original installation — four double pipe sections 2 parallel × 2 series

**Figure 2b** — to increase capacity 50% — merely add one bank 3 parallel × 2 series

**Figure 2c** — To decrease capacity — shut off one bank 1 parallel × 2 series

**Figure 2** — How hairpin sections are changed to meet new requirements

**Figure 2.36** Double-pipe hairpin banks.

**Figure 2.37** Stacked double-pipe exchanger.

Can handle up to 10 fluids in a single exchanger "cold box"

Composed of alternating layers of corrugated fins and flat
separator sheets called parting sheets

A stack of fins and parting sheets make up this heat
exchanger

Fins are constructed in four pattern types (Figure 2.39)
    Plain
    Perforated
    Herringbone
    Sherrated

**Figure 2.38** Vacuum-brazing process.

PLAIN  • A sheet of metal with corrugated fins at right angles to the plates.

PERFORATED • A plain fin constructed from perforated material.

HERRINGBONE • Made by displacing the fins sideways at regular intervals to produce a zig-zag effect.

SERRATED • Made by simultaneously folding and cutting alternative sections of fins. These fins are also known as the lanced or multi-entry pattern.

**Figure 2.39** Fin patterns for a brazed aluminum plate-fin exchanger.

Figure 2.40 Brazed aluminum plate-fin exchangers.

### Dependent on Application Requirements

Number of layers

Types of fins stacking arrangement

Stream circuiting will vary depending on application requirements

**Fluid configurations include** cross-flow, counterflow, and cross-counterflow

See also Figures 2.40–2.54.

### ▶ PLATE AND FRAME EXCHANGERS

### Overview

Consists of layers of thin ribbed alloy plates, spaced about ¼-in. apart, compressed together between gaskets (see Figure 2.55).

**Figure 2.41** Brazed aluminum plate-fin exchangers.

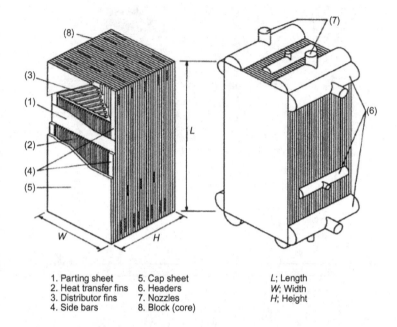

1. Parting sheet
2. Heat transfer fins
3. Distributor fins
4. Side bars
5. Cap sheet
6. Headers
7. Nozzles
8. Block (core)

L; Length
W; Width
H; Height

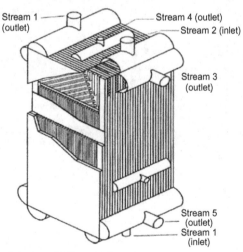

**Figure 2.42** Brazed aluminum plate-fin construction.

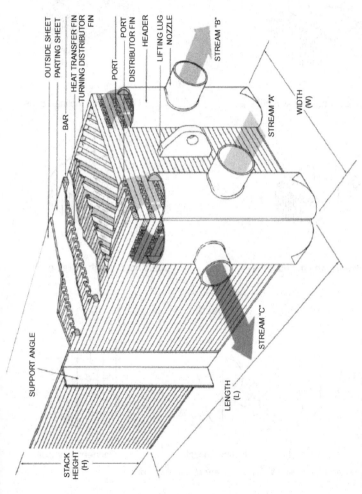

**Figure 2.43** Brazed aluminum plate-fin construction.

Standard header with flat ends

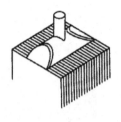

Header with inclined ends

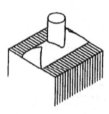

Header with mitred ends

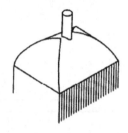

Dome header with mitred ends

Radial nozzle

Inclined nozzle

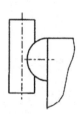

Tangential nozzle

**Figure 2.44** Brazed aluminum plate-fin exchanger construction—Headers.

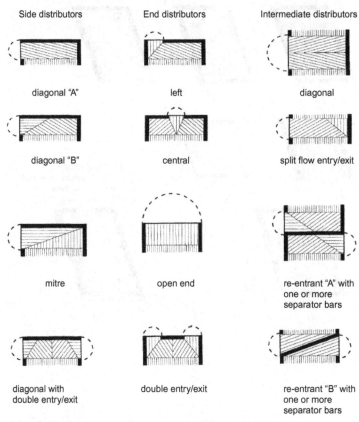

**Figure 2.45** Brazed aluminum plate-fin construction—Distributors.

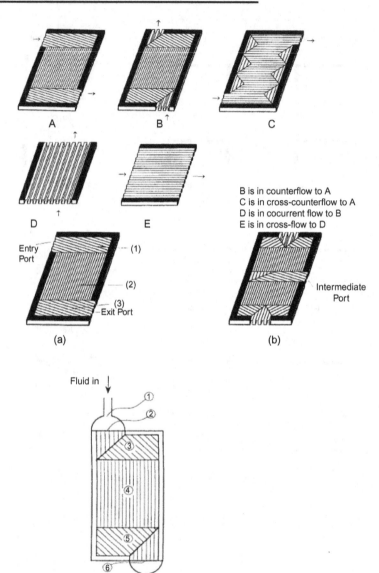

B is in counterflow to A
C is in cross-counterflow to A
D is in cocurrent flow to B
E is in cross-flow to D

Figure 2.46 Brazed aluminum plate-fin construction—Layout.

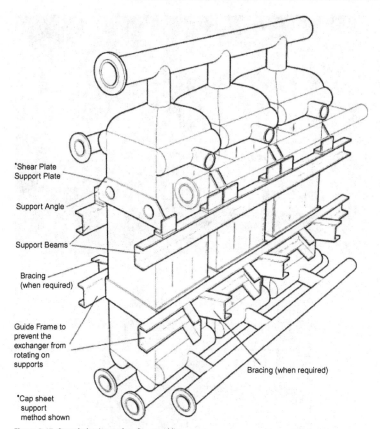

*Shear Plate
Support Plate

Support Angle

Support Beams

Bracing
(when required)

Guide Frame to
prevent the
exchanger from
rotating on
supports

*Cap sheet
support
method shown

Bracing (when required)

**Figure 2.47** Brazed aluminum plate-fin assemblies.

Figure 2.48 Brazed aluminum plate-fin assemblies.

**Figure 2.49** Brazed aluminum plate-fin pressure vessels.

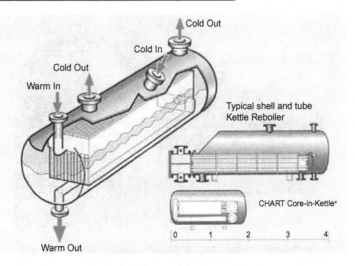

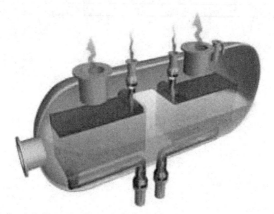

**Figure 2.50** Brazed aluminum plate-fin core-in kettles.

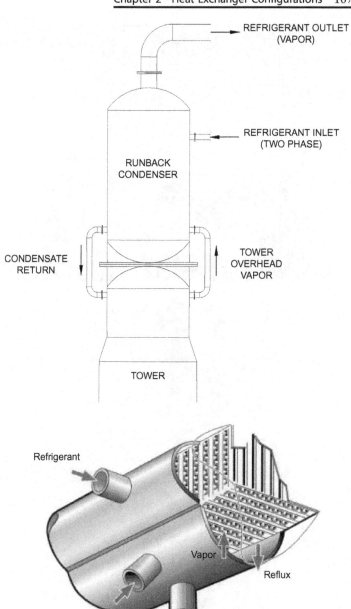

**Figure 2.51** Brazed aluminum plate-fin reflux condensers.

STANDARD TRANSITION JOINT

FLANGE REINFORCED TRANSITION JOINT

**Figure 2.52** Brazed aluminum plate fin in cold boxes.

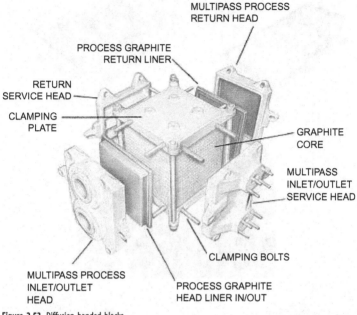

**Figure 2.53** Diffusion bonded blocks.

**Figure 2.54** Plate-fin tube coil configuration.

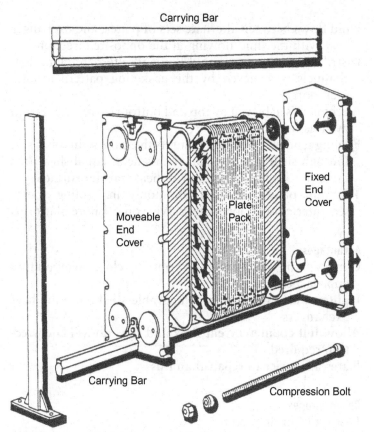

**Figure 2.55** Plate-and-frame exchanger.

Fluid flows between alternate sets of plates, and exchanges heat with the fluid flowing in the opposite direction

Distribution of hot and cold fluids to alternate plate flow channels is achieved by the gasketing pattern around each port

Alignment is achieved by top and bottom carrying bars and by slots in each plate

Fouling tendency between plates is the same as in tubes and through shell side at equal shell side at equal shear (for same pressure drop per unit of heat transfer surface)

Titanium plates are used commonly in cooling water/seawater/closed loop water service on offshore platforms

## Advantages

Compact in size and can obtain **close temperature approaches**

**Lighter and smaller** than comparable shell-and-tube heat exchangers

Allow **full countercurrent flow** and thus **no MTD correction required**

Plates are **easier to repair** than fins

## Disadvantages

Less efficient, limited to

Moderate temperature [$< 250°$F ($< 121°$C)]

Moderate pressure [$< 200$ psig ($< 13.79$ barg)] due to gasketed service

Should be used only where excessive fouling is not expected

Requires superior metallurgy

Gaskets can deteriorate in hydrocarbon service, poor fire resistance, and expensive

Arrangement of gasketed, pressed metal plates aligned on carrying bars and secured between two covers by compression bolts (Figure 2.56).

Provides increased surface area

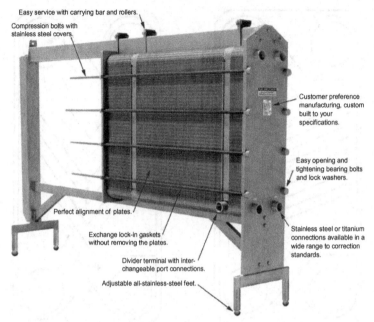

**Figure 2.56** Gasketed plate-and-frame exchanger.

Plates are made of thin pressed metal that are resistant to corrosive attack.

SS, Monel, titanium, aluminum, bronze
Attractive to seawater/brackish services
Gasket material must be compatible with service

## Plates
May be high theta or low theta plates
High theta plates offer
 Better heat transfer
 Higher pressure drop
Multiple plate styles may be present in a single unit
Different combinations of plates yield different results
 when used together

See also Figures 2.57–2.68.

**Figure 2.57** Traditional gasketed plate-and-frame design.

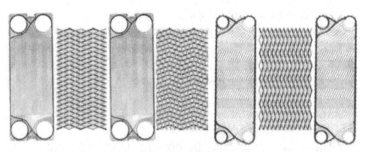

**Figure 2.58** Multiple plate styles in a traditional plate-and-frame exchanger.

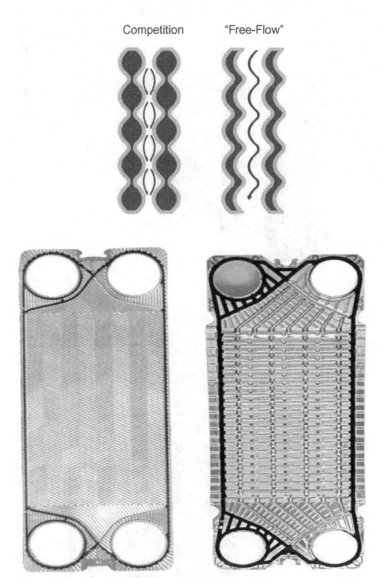

**Figure 2.59** Traditional gasketed plate and frame with wide gap plate design.

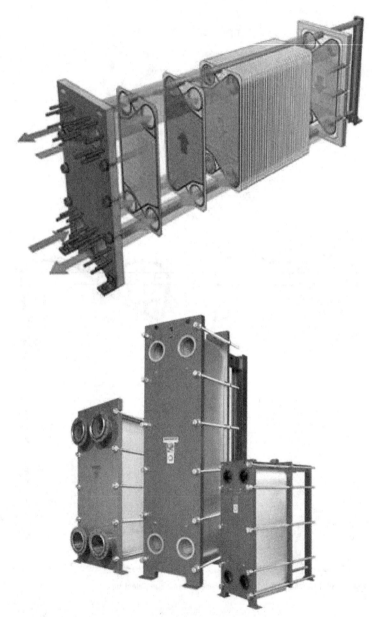

**Figure 2.60** Traditional gasketed plate-and-frame design.

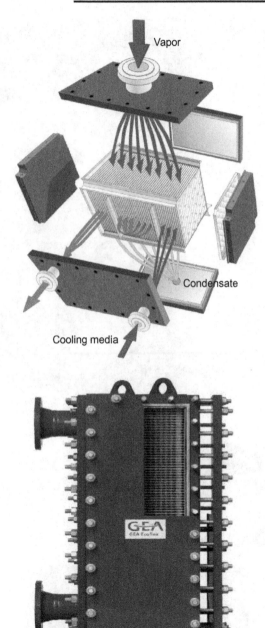

Figure 2.61  Welded "printed circuit" plate and frame design.

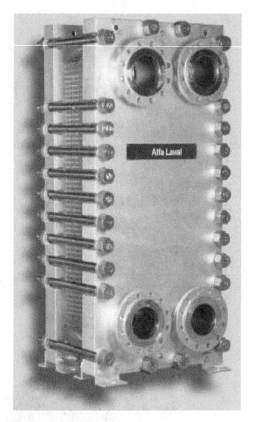

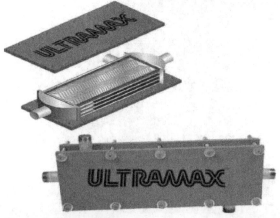

Figure 2.62  Welded plate and frame

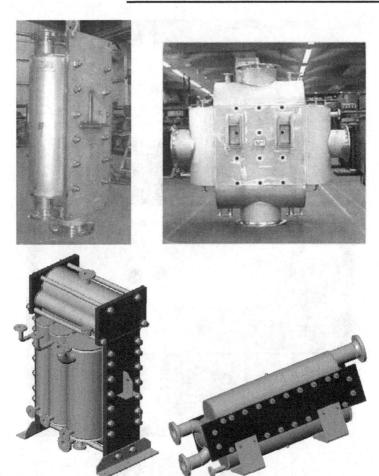

**Figure 2.63** Welded plate-and-frame hybrid design.

**Figure 2.64** Welded plate-and-frame exchangers.

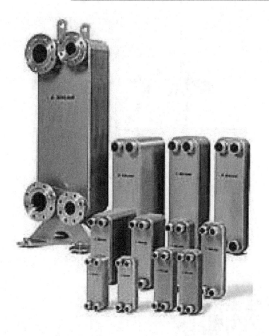

Figure 2.65 Brazed plate-and-frame exchanger design.

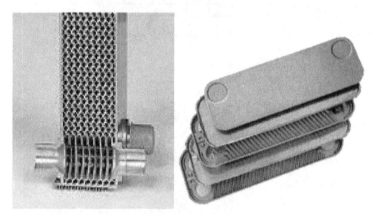

**Figure 2.66** Brazed plate-and-frame exchanger design.

**Compact doesn't necessarily mean small!**

**Figure 2.67** Compact doesn't necessarily mean small! Welded plate-and-frame exchanger (Alfa Laval/ Packinox).

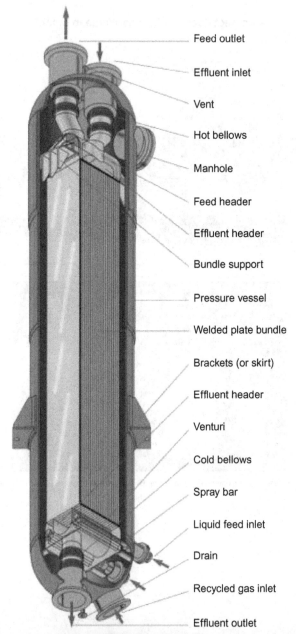

Feed outlet

Effluent inlet

Vent

Hot bellows

Manhole

Feed header

Effluent header

Bundle support

Pressure vessel

Welded plate bundle

Brackets (or skirt)

Effluent header

Venturi

Cold bellows

Spray bar

Liquid feed inlet

Drain

Recycled gas inlet

Effluent outlet

Figure 2.68  Welded plate-and-frame exchanger (Alfa Laval/Packinox).

## ▶ INDIRECT-FIRED HEATERS

Hot combustion gas and flame heat an intermediate liquid, which, in turn, heats a fluid flowing through a coil or a series of tubes

### Advantages

Maintains a constant temperature over a long period of time

Has proven safe, reliable, and convenient to use

### Disadvantage

Requires several hours to reach the desired temperature after it has been out of service

### Intermediate Liquid

Must be stable at atmospheric pressure and maximum temperature involved

May be (depending on temperature level required)
  Water
  Heating fluid medium

Transfer heat between the fire tube and the fluid being heated by natural convection
  This limits the rate of heat flux per unit area
  Thus, seldom used to produce outlet temperatures above 500°F

Used primarily to heat oil and gas in production operations where heat loads are not large, for example,
  Line heaters in natural gas service (Figure 2.69)
  Reboilers in amine (Figure 2.70)

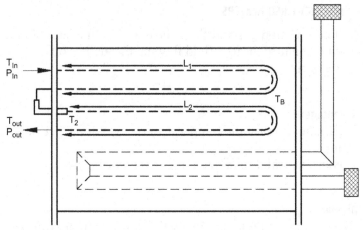

**Figure 2.69** Schematic of a line heater.

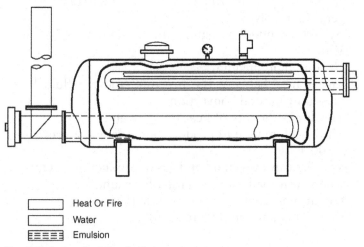

Heat Or Fire
Water
Emulsion

**Figure 2.70** Cutaway of an indirect-fired heater.

Glycol and oil stabilization units (the intermediate liquid is being boiled, thus there is no fluid coil or tube)

Units usually consist of (Figure 2.71)

Fire tube (usually fired by gas)

Coil for fluid to be heated (Figure 2.72)

Immersed in a heat transfer fluid

Designer must specify

Heat duty

Size of fire tube

Coil diameter, length, and wall thickness

## Sizing Considerations

To determine **heat duty** required, we must know

**Maximum** gas, water, and oil/condensate **expected** in the heater

Pressure and temperatures of heater **inlet and outlet**

Outlet temperature on gas "line" heaters depends on hydrate formation temperature of the gas

Coil size of heater depends on volume of fluid flowing through the coil and required heat duty

Special operating conditions must be considered in sizing

**Start-up** of a **shut-in** well may require **additional** heater **capacity** over steady-state requirements

## Heat Duty

Heat duty should be **checked** for **various combinations** of inlet temperature, pressure, flow rate, and outlet temperature and pressure so as to determine **most critical combination**

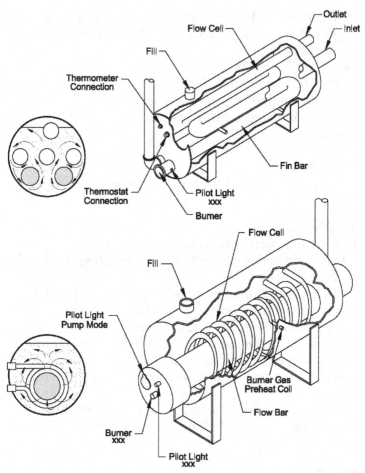

**Figure 2.71** Types of indirect heaters.

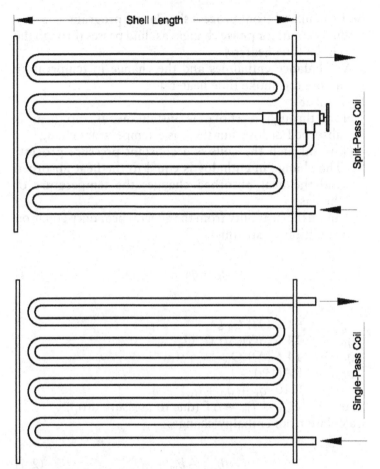

**Figure 2.72** Indirect heating coils.

Best calculated from **process simulation program**

Will account for **phase changes** as fluid passes through the choke (line heaters)

Will **balance enthalpies** and the **change** in **temperature** across the choke (line heaters)

Remember

Flow through a choke is **instantaneous,** no heat is absorbed or lost, but there is a temperature change

Flow through the **coils** is a **constant pressure process. The change in enthalpy is equal** to the **heat absorbed** and the **heat absorbed changes** the **temperature** of the gas

Calculate the heat duty from the general heat duty equation for multiphase streams

$$q = q_g + q_0 + q_w + q_l \qquad (2.3)$$

where

$$
\begin{aligned}
q_g &= 41.7\,(\Delta t)\,C_g\,Q_g \\
q_0 &= 14.6\,(SG)(\Delta t)\,C_g\,Q_g \\
q_w &= 14.6\,(\Delta t)\,Q_w \\
q_{lost} &= UA(190 - t_{amb}) \\
&= 0.10\,(q_g + q_0 + q_w) \\
\Delta t &= T_{out} - T_{in} + \Delta T \text{ (due to pressure drop)}
\end{aligned}
$$

Calculate from enthalpy changes

$$q = h_{out} - h_{in} + q_{lost} \qquad (2.4)$$

where

$$
\begin{aligned}
h_{out} &= \text{enthalpy at } P_{out},\ T_{out} \\
h_{in} &= \text{enthalpy at } P_{in},\ T_{in}
\end{aligned}
$$

## Sizing Fire Tubes

Calculate **area of fire tube** with a **flux rate of 10,000 Btu/hr ft$^2$** (recommended heat flux for maximum life)

$$L = 3.8x10^{-4}\left(\frac{q}{d}\right) \qquad (2.5)$$

where
$L$ = fire tube length, ft.
$q$ = total heat duty, Btu/hr
$d$ = fire tube diameter, in.
Burner should be selected from standard off-the-shelf units
Manufacturers normally have standard diameters and lengths for different size fire tube ratings (Table 2.7)
In order to choose the coil length and diameter, a temperature must first be chosen upstream of the choke

## Coil Sizing

**Choose temperatures** (upstream of choke)
The greater the temperature difference between the gas and the bath, the smaller the coil area required

**TABLE 2.7** Standard Size Fire Tubes for Indirect-Fired Heaters

| | |
|---|---|
| 250,000 Btu/hr | 4,000,000 Btu/hr |
| 500,000 Btu/hr | 4,500,000 Btu/hr |
| 750,000 Btu/hr | 5,000,000 Btu/hr |
| 1,000,000 Btu/hr | 6,000,000 Btu/hr |
| 1,500,000 Btu/hr | 7,000,000 Btu/hr |
| 2,000,000 Btu/hr | 8,000,000 Btu/hr |
| 2,500,000 Btu/hr | 9,000,000 Btu/hr |
| 3,000,000 Btu/hr | 10,000,000 Btu/hr |
| 3,500,000 Btu/hr | |

Bath temperature is constant (use 190°F)

Gas will be coldest downstream of choke

Therefore, the shortest total coil length ($L_1 + L_2$) will occur when $L_1$ is small

If $L_1$ becomes too long, consider using a glycol/water mixture or other heating medium or raise the bath temperature

Try and keep $T_2$ above 50°F so as to minimize plugging and above −20°F to avoid more costly steel for sure

Choose coil diameter

Erosional flow will **always govern** when choosing the diameter

$$V_e = \frac{c}{\rho_m^{1/2}} \qquad (2.6)$$

Normally 60 fps

$CO_2$ service limit velocity to 50 fps

Pressure drop usually does not govern

Wall thickness determination

ANSI B31.8, B31.3, ASME Pressure Vessel Code, API 12K

**Before** choosing wall thickness, select **pressure rating of coil**

High-pressure coil—SITP

Low-pressure coil—Maximum allowable working pressure of downstream equipment

Calculate heat duty ($q$) for each coil

Calculate

$$MTD = \frac{\Delta t_1 - \Delta t_2}{l_{N_e}\left(\dfrac{\Delta t_1}{\Delta t_2}\right)} \qquad (2.7)$$

where

$\Delta t_1$ = temperature difference between coil inlet and bath

$\Delta t_2$ = temperature difference between coil outlet and bath

**Calculate** U or use **appropriate table** or chart (typical range 44–53 Btu/hr ft$^2$ °F)

**Calculate coil length**

Because $U$, $MTD$, $Q$, and the diameter of the pipe are known, the length of the coil can be calculated

$$L = \frac{12q}{\pi(MTD)Ud} \qquad (2.8)$$

where

$d$ = coil diameter, in.

**Equation (2.6)** describes an **overall length** required for the coil

### Heater Sizing

Based on previous information

Choose shell length and number of passes

As shell length decreases, the number of passes increases and a larger diameter is required

The preceding procedure is helpful for **reviewing existing designs** or **vendor proposals**

However, it is more **economically advantageous** to select **standard size heaters** from the **manufacturer**

See Figure 2.73.

### ▶ DIRECT-FIRED HEATERS

Used when large amounts of heat input are required

Consists of two types

Those that heat the fluid with a fire tube directly

Those that heat the fluid using both radiant and convection

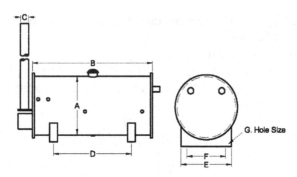

## NOMINAL DIMENSIONAL DATA

| Heater | A | B | C | D | E | F | G |
|---|---|---|---|---|---|---|---|
| BTU HR | Ft. In. | Ft. In. | Ft. In. | Ft. In. | Ft. In. | Ft. In. | In. only |
| 250,000 | 2'-0" | 7'-6" | 0'-8" | 5'-6" | 1'-0" | 1'-5" | 3/4" |
| 500,000 | 2'-6" | 10'-0" | 0'-10" | 6'-0" | 1'-9" | 1'-5" | 11/16" |
| 750,000 | 3'-0" | 12'-0" | 0'-12" | 6'-0" | 2'-2" | 1'-10" | 11/16" |
| 1,000,000 | 3'-6" | 14'-4" | 1'-2" | 11'-0" | 3'-0" | 2'-4" | 3/4" |
| 1,500,000 | 4'-0" | 17'-6" | 1'-4" | 12'-6" | 3'-6" | 3'-0" | 3/4" |
| 2,000,000 | 5'-0" | 20'-0" | 1'-8" | 12'-6" | 4'-4" | 3'-0" | 7/8" |

## SPECIFICATIONS

| Heater Furnace Input BTU/HR | Shell Size O.D. x Lgt. | Std. No. & Size Tubes | Coil W.P. PSI | Std. Mean Coil Area Sq. Ft. | Approx. Coil. Lin. Ft. | Water Fill Vol: Bbls. | Shipping Weight Pounds |
|---|---|---|---|---|---|---|---|
| 250,000 | 24"x7' 6" | 8-2"XH | 3372 | 29.5 | 54 | 2.9 | 1,400 |
| 250,000 | 24"x7' 6" | 8-2"XXH | 6747 | 26.5 | 54 | 2.9 | 1,610 |
| 500,000 | 30"x10' 0" | 8-2"XH | 3372 | 42.6 | 76 | 6.0 | 2,210 |
| 500,000 | 30"x10' 0" | 8-2"XXH | 6747 | 38.3 | 76 | 6.0 | 2,510 |
| 750,000 | 36"x12' 0" | 10-2"XH | 3372 | 64.4 | 114 | 10.5 | 2,875 |
| 750,000 | 36"x12' 0" | 10-2"XXH | 6747 | 58.8 | 114 | 10.5 | 3,325 |
| 750,000 | 36"x12' 0" | 6-3"XH | 3150 | 59.4 | 70.9 | 10.3 | 3,030 |
| 750,000 | 36"x12' 0" | 6-3"XXH | 6300 | 58.8 | 70.9 | 10.3 | 3,615 |
| 1,000,000 | 42"x14' 4" | 12-2"XH | 3372 | 93.4 | 166 | 17.9 | 4,060 |
| 1,000,000 | 42"x14' 4" | 12-2"XXH | 6747 | 85.9 | 166 | 17.9 | 4,725 |
| 1,000,000 | 42"x14' 4" | 8-3"XH | 3150 | 94.8 | 113.2 | 17.5 | 4,390 |
| 1,000,000 | 42"x14' 4" | 8-3"XXH | 6300 | 85.9 | 113.2 | 17.5 | 5,335 |
| 1,500,000 | 48"x17' 6" | 14-2"XH | 3372 | 134.0 | 237 | 28.7 | 5,650 |
| 1,500,000 | 48"x17' 6" | 14-2"XXH | 6747 | 120.5 | 237 | 28.7 | 6,600 |
| 1,500,000 | 48"x17' 6" | 10-3"XH | 3150 | 145.0 | 173.1 | 28.0 | 6,235 |
| 1,500,000 | 48"x17' 6" | 10-3"XXH | 6300 | 131.4 | 173.1 | 28.0 | 7,675 |
| 2,000,000 | 60"x20' 0" | 16-2"XH | 3372 | 175.7 | 311 | 51.8 | 10,110 |
| 2,000,000 | 60"x20' 0" | 16-2"XXH | 6747 | 158.0 | 311 | 51.8 | 11,360 |
| 2,000,000 | 60"x20' 0" | 10-3"XH | 3150 | 165.9 | 198.1 | 51.2 | 10,580 |
| 2,000,000 | 60"x20' 0" | 10-3"XXH | 6300 | 150.4 | 198.1 | 51.2 | 12,240 |

Figure 2.73 Standard indirect heaters and coil sizes.

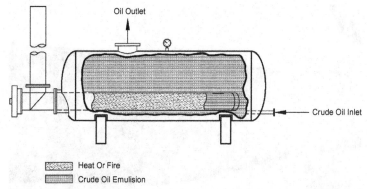

**Figure 2.74** Cutaway of a direct-fired heater.

Direct-fired heaters using a fire tube to heat the fluid directly (Figure 2.74)

    Oil passes through an inlet distributor and is heated directly by a fire box

    Quick, efficient (75 to 90%), and the initial cost is relatively low

    If fuel gas is available, utilization of a direct-fired heater for oil treating (especially high-volume oil treating) should be considered

    Hazardous and require special safety equipment

    Scale may form on the oil side of the fire tube, preventing the transfer of heat from the fire box to the oil emulsion

        Heat collects in the steel walls under the scale

        Metal softens and buckles (Figures 2.75 and 2.76)

        Metal eventually ruptures, allowing oil to flow into the fire box

        Resultant blaze, if not extinguished, will be fed by the incoming oil stream

**Figure 2.75** Fire tube scale prevents heat transfer.

Direct-fired heaters that heat fluid using both radiant and
convection heating
  Because the temperature in both sections is very high, an
    intermediate heat transfer fluid is typically used
  Wide variety of heating configurations; choice depends on
    Fuel cost
    Thermal efficiency
    Temperature desired
    Size of the heat load
    Fluid being heated
  Two basic configurations
    Horizontal tubes
    Vertical tubes

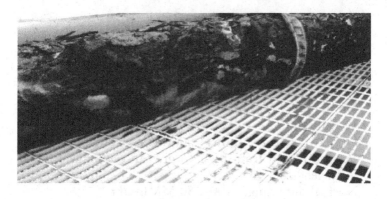

**Figure 2.76** Fire tube scale and potential hot spots.

**Horizontal Tubes** (Refer to Figure 2.77)

Cabin
  Radiant section normally lines the walls with burners in the
    floor
  Economical, efficient, and most popular
  Normal duty range: 10–100 MMBtu/hr
Two-cell box
  Only two cells are shown, but three or four can be used
  Vertically fired from floor to give an economical, high-
    efficiency design
  Normal duty range; 100–250 MMBtu/hr
Cabin with bridgewall
  Provides two sections that can be fired individually
  Can be fired horizontally or vertically
  Normal duty range: 20–100 MMBtu/hr

**End Fired Box**

Horizontally fired as name implies
  Normal duty range: 5–50 MMBtu/hr
End fired box with side-mounted convection section
  Older type unit that may be used in new installations with
    high ash, poor grade fuels
  More expensive design
  Normal duty range: 50–200 MMbtu/hr
Single row, double fired
  Consists of one or more cells shown
  Used for reactor-fed heating units
  Normal duty range: 20–50 MMbtu/hr

**Vertical Tubes** (Refer to Figure 2.78)

All radiant
  Low-cost, low-efficiency design that is compact
  Normal duty range: 0.5–20 MMbtu/hr

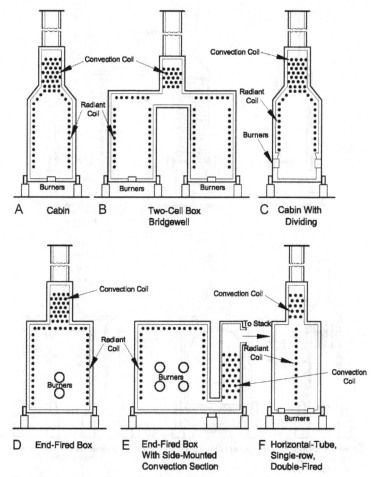

**Figure 2.77** Basic types of direct-fired heaters with horizontal tube.

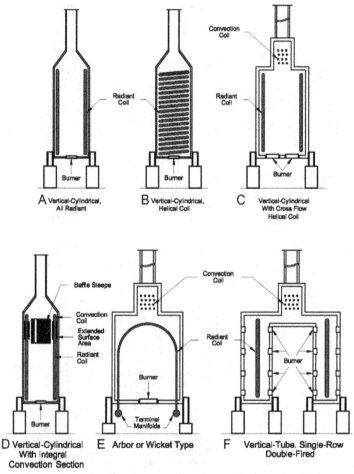

**Figure 2.78** Basic types of direct-fired heaters with vertical tube red heaters.

Cylindrical, helical coil
  Basic low-cost, low-efficiency alternate to the "all radiant" type
  Not feasible to have parallel flow coils for fluid
  Normal duty range: 0.5–20 MMbtu/hr
Cylindrical, with cross-flow convection
  Popular new vertical flow units
  Economical, efficient, and compact
  Normal duty range: 10–200 MMbtu/hr
Cylindrical, with integral convention
  Many existing units in service
  Not commonly used due to its limited thermal efficiency
  Normal duty range: 0–100 MMbtu/hr
Arbor or wicket
  Used to heat large quantities of gas where low pressure drop is desired
  Several arbor coils may be used in one heater unit
  Normal duty range: 50–100 MMbtu/hr
Single-row, double-fired
  Most expensive configuration but provides high and rather uniform heat flux
  Normal duty range: 20–125 MMbtu/hr

### Development of "Hot Spots" and Tube Failure Result from
High flame temperatures
Low convection film coefficient
Tube metallurgy selection is a compromise between initial cost and service life
Choice of material, method of welding, configuration used, and so on must be based on experience

### Thermal Efficiency Considerations
Control of excess air
Additional convection heat recovery (Figure 2.79)
Preheating the combustion air

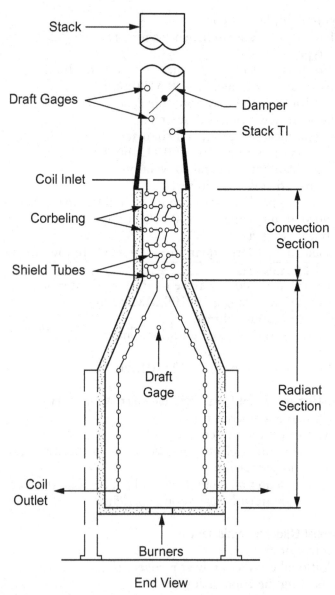

**Figure 2.79** Fired heater.

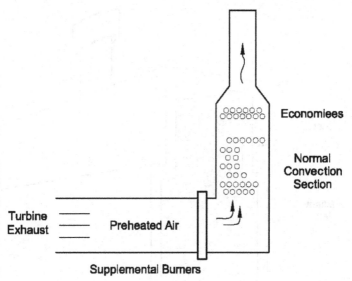

**Figure 2.80** Waste heat recovery.

Using turbine exhaust gas (waste heat) (Figure 2.80)

Excess air requires fuel to heat up ambient air to the stack temperature

Thermal efficiency is inversely proportional to excess air; stack air varies directly

Use of studded or fin type convection tubes has enhanced heat recovery from the hot combustion gases (Figure 2.81)

Use of extended surface on the low coefficient side tends to reduce fouling inside the tubes, a possible major problem. However, said surfaces foul easier

Combustion air preheat with combustion gas is common

Thermal efficiency may reach 90% with the stack temperature lowered to 300–350°F

Tends to elevate combustion temperatures, which increase radiant heat transmission and tube surface temperature

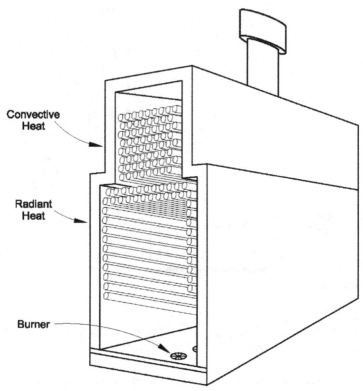

**Figure 2.81** Direct-fired furnace.

Gas turbine exhaust can be a super substitute for air
   Contains 17–18% oxygen
   Already preheated to 800–900°F
   Combustion gas is flowed through a convection section to
     heat process streams
   This heat is used directly or indirectly to generate steam,
     run boilers, regenerate solid desiccant units, heat gas
     and liquid streams, etc.

**Determining Required Heat Input**

General heat transfer equation

Expressed by the following equation:

$$q = WC_p\Delta T \qquad (2.9)$$

where

$q$ = heat required, Btu/hr

$W$ = flow rate, lb/hr

$C_p$ = specific heat constant, Btu/lb °F

= 1.0 for water

= 0.5 for oil (average)

$\Delta T$ = differential temperature, °F

Flow rate determination

Because **water weighs 350 lbs/bbl,** the flow rate can be expressed as follows employing conversion factors

$$W = \left(\frac{350}{24}\right)(SG)_L Q_L \qquad (2.10)$$

where

$(SG)_L$ = specific gravity of liquid

$Q_L$ = liquid flow rate, BPD

Total heat required

Expressed from the following equation:

$$q = q_0 + q_w + q_g + q_{lost} \qquad (2.11)$$

where

$q_0$ = heat required to heat the oil

= $[(350/24)(SG)_0 \, Q_0](0.5) \, \Delta T$

$q_w$ = heat required to heat the water

= $[(350/24)(SG)_w \, Q_0](1.0) \, \Delta T$

$q_g$ = 0

Substituting into Equation (2.11)

$$q = \frac{350}{24}\left[(SG)_0 Q_0(.5) + (SG)_w Q_w(1)\right]\Delta T + q_{lost} \quad (2.12)$$

Assuming that the **free water** has been **removed** so that $Q_w = 0.1\, Q_0$, $(SG)_W = 1.0$, and $q_{lost} = (0.1)(q)$, then the following equation exists:

$$q = 15\left[(0.5)(SG)_0 Q_0 + (SG)_w Q_w\right]\Delta T + q_{lost} \quad (2.13)$$

Simplifying

$$q = 16 Q_0 \Delta T\left[(0.5)(SG)_0 + 0.1\right] \quad (2.14)$$

## ▶ AIR-COOLED EXCHANGERS

### General Considerations

Expensive but cost-efficient

Cool or condensate low-viscosity fluids

Usually designed by the manufacturer

Air cooling is **preferred** if the ambient air temperature is **low enough** to provide **efficient cooling**

**Cooling towers** have no **application** in offshore, **arctic,** or humid locations

Popular even where water for cooling is available because
  Mechanically simple
  Flexible
  Eliminates the nuisance and cost of water treating
In warm climates
  May not be capable of producing as low a temperature as water
  Study should be conducted of cooling alternatives

Common uses
  Cool hot fluid to something near ambient
  Interstage cooler in compression

## Typical Air-Cooled Exchanger Configurations
Four rows of tubes on staggered pitch for moderate temperature ranges
Six rows of tubes for a large temperature range service
Pitch of 2½ inch (63.5 mm) with equilateral triangular layout
1-in. OD tubes with ten 5/8-in. (15.875-mm) high aluminum fins per inch
20:1 finned to outside bare surface area ratio
30-ft (9144 mm) tube length (when mounted over pipe way)
At least two fans per bay
  Half are autovariable pitch
Aluminum or plastic fan blades
Draft types
  Forced draft
    Less expensive
    Tube bundle located on discharge of fan
    Used when outlet air is too hot for fan parts
  Induced draft
    More efficient
    Tube bundle located on suction of fan
    Used when fans off-performance is needed
Multiple services (bundles) located in same bay
  Caution of "overcooling"
Winterization
  Warm air recirculation with steam

## Advantages of Forced Draft Design
Easy to remove and replace bundles

Easier to mount motors or other drivers with short shafts

Lubrication, maintenance, etc., more accessible

With reinforced straight side panels to form a rectangular box type of plenum, shipping and mounting are simplified greatly, thus permitting complete preassembled shop-tested units

Best adapted for cold climate operation with warm air recirculation

Requires less horsepower for an air temperature rise greater than 50°F

### Disadvantages of Forced Draft Design

Possibility that hot air leaving the top will flow around the unit and be drawn through again

Less efficient air distribution over the bundle

Harder to clean when covered with lint, bugs, or debris

Hoods do not offer protection from weather

### Advantages of Induced Draft Design

Easier to shop assemble, ship, and install

Hoods offer protection from weather

Easier to clean underside when covered with lint, bugs, or debris

More efficient air distribution over the bundle

Less likely to be affected by hot air recirculation

Requires less horsepower for an air temperature rise less than 50°F

### Disadvantages of Induced Draft Design

More difficult to remove bundles for maintenance

High temperature service limited due to effect of hot air on fans

More difficult to work on fan assembly, i.e., adjust blades due to heat from bundle, and their location

| Tube lengths | - | 6 to 50 ft |
| Tube diameter | - | 5/8 to 1½ in. |
| | - | Air is nonfouling and offers low heat transfer efficiency |
| Fins | - | ½- to 1-in. height |
| | - | 7 to 11 per inch |
| Depth | - | 3 to 8 rows of fin tubes |
| | - | Triangular pitch with fins separated by 1/16 to ¼ in. |
| Bay widths | - | 4 to 30 ft |
| Fan diameters | - | 3 to 16 ft |

See Figures 2.82–2.91.

### Air-Side Control

Adjustment must be made to **assure** adequate cooling while not overcooling process

  Process flow rate/heat duties change

  Air temperatures change

    Season to season

    Night to day

Too cool gas temperature could lead to hydrate formation

Too cool lube oil temperature leads to high viscosity, which results in high $\Delta p$ and poor lubrication

Process outlet temperature is controlled by

  **Louvers** (most common)

    Energy inefficient

    **Mechanical** (seasonal or night/day air temperature changes)

    Automatic (senses process temperature)

  Variable pitch fan blades (second most common)

  Variable speed drivers (third most common)

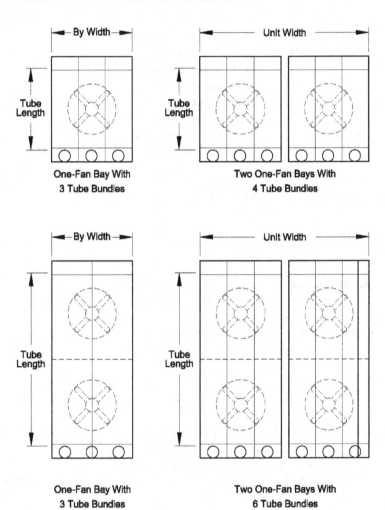

**Figure 2.82** Air-cooled exchanger bundles and bays.

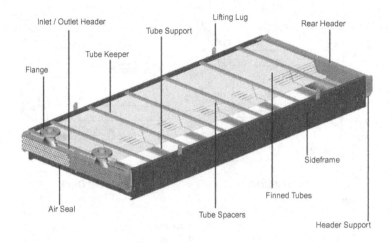

Inlet / Outlet Header

Tube Support

Lifting Lug

Rear Header

Tube Keeper

Flange

Air Seal

Tube Spacers

Finned Tubes

Sideframe

Header Support

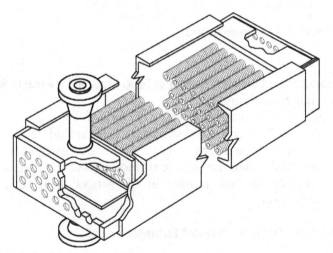

**Figure 2.83** Air-cooled exchanger construction.

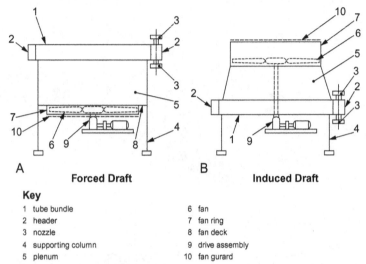

**A**   Forced Draft          **B**   Induced Draft

**Key**

| | | | |
|---|---|---|---|
| 1 | tube bundle | 6 | fan |
| 2 | header | 7 | fan ring |
| 3 | nozzle | 8 | fan deck |
| 4 | supporting column | 9 | drive assembly |
| 5 | plenum | 10 | fan gurard |

**Figure 2.84** Air-cooled exchanger plug header.

### Procedure for Calculating Number of Tubes Required for Aerial Coolers

Similar to shell and tube

Figure 2.92 lists **approximate "U" values**

Use "$U_x$" when **extended surface area,** including fins, is used for the **area term** in the general heat transfer equation

Use "$U_b$" when the **outside surface area** of the bare tube (neglecting fins) is used in the general heat transfer equation

### Procedure for Sizing Air-Cooled Exchanger

Use "$U_x$"

Approximate air temperature rise (exit air temperature)

$$\Delta t_a = \left(\frac{U_x + 1}{10}\right)\left(\frac{t_1 + t_2}{2} - t_1\right) \qquad (2.15)$$

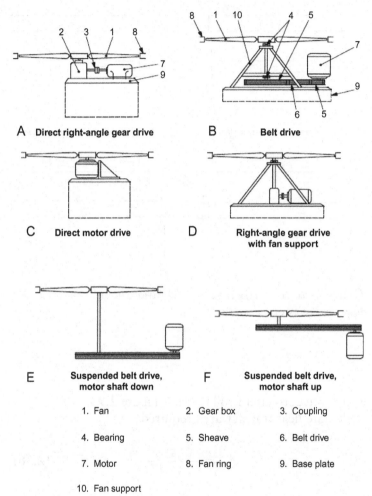

**A**  Direct right-angle gear drive

**B**  Belt drive

**C**  Direct motor drive

**D**  Right-angle gear drive
   with fan support

**E**  Suspended belt drive,
   motor shaft down

**F**  Suspended belt drive,
   motor shaft up

| | | |
|---|---|---|
| 1. Fan | 2. Gear box | 3. Coupling |
| 4. Bearing | 5. Sheave | 6. Belt drive |
| 7. Motor | 8. Fan ring | 9. Base plate |
| 10. Fan support | | |

**Figure 2.85** Air-cooled exchanger drives.

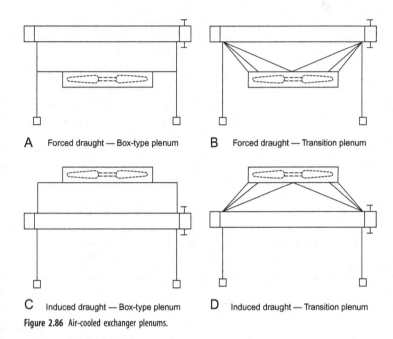

A   Forced draught — Box-type plenum       B   Forced draught — Transition plenum

C   Induced draught — Box-type plenum       D   Induced draught — Transition plenum

**Figure 2.86** Air-cooled exchanger plenums.

Calculate corrected LMTD using Figure 2.93.
Calculate heat transfer area required

$$A_x = \frac{Q}{U_x\, LMTD} \tag{2.16}$$

$$F_a = \frac{A}{APSF} \tag{2.17}$$

Calculate bundle face area from Table 2.8.

**Figure 2.87** Air-cooled exchanger fans.

Calculate width from assumed length of tubes

$$Width = \frac{F_a}{L} \qquad (2.18)$$

Calculate number of tubes

$$N_t = \frac{A_x}{(APSF)L} \qquad (2.19)$$

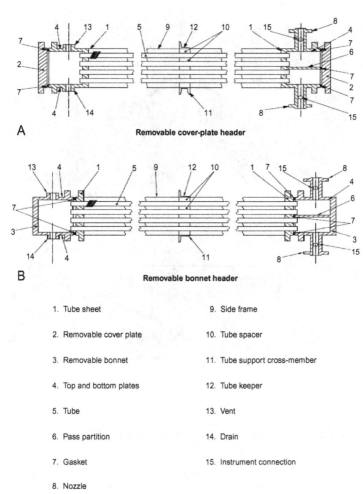

**A**   Removable cover-plate header

**B**   Removable bonnet header

1. Tube sheet

2. Removable cover plate

3. Removable bonnet

4. Top and bottom plates

5. Tube

6. Pass partition

7. Gasket

8. Nozzle

9. Side frame

10. Tube spacer

11. Tube support cross-member

12. Tube keeper

13. Vent

14. Drain

15. Instrument connection

**Figure 2.88** Air-cooled exchanger cover-plate header.

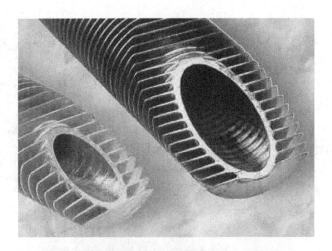

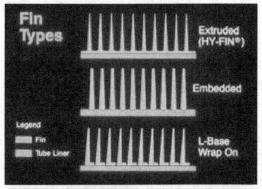

Cutaway view of fin tubes

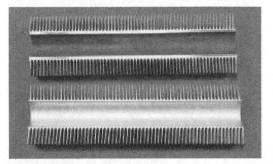

**Figure 2.89** Cutaway of finned tubes.

**Figure 2.90** Air-cooled exchangers.

**Figure 2.91** Air-cooled exchanger.

Tables, graphs, and procedures are given in "GPSA Engineering Data" book to calculate
Pressure drop in tubes
Actual $U_x$ to compare to assumed
Size, number, and horsepower of fans required
Air heat effectiveness compared to water
2% without fins
20% with fins on the air side (ignoring fouling effect of water on exchanger)
50–70% with the type of water used

### When Designing We Must Consider
Cooling tube design
Fan arrangement
Number of transfer bays

| | | Finetube | | | |
|---|---|---|---|---|---|
| Service | | ½ in. By 9 | | 5/8 in. By 10 | |
| Water & Water Solutions | | | | | |
| | | $U_b$ | $U_x$ | $U_b$ | $U_x$ |
| Engine jacket water (rf = .001) | | 110 | – 7.5 | 130 | – 6.1 |
| Process water (rf = .002) | | 95 | – 6.5 | 110 | – 5.2 |
| 50-50 ethyl glycol-water (rf = .001) | | 90 | – 6.2 | 105 | – 4.9 |
| 50-50 ethyl glycol-water (rf = .002) | | 80 | – 5.5 | 95 | – 4.4 |
| Hydrocarbon Liquid Coolers | | | | | |
| | Viscosity Cp | $U_b$ | $U_x$ | $U_b$ | $U_x$ |
| | 0.2 | 85 | – 5.9 | 100 | – 4.7 |
| | 0.5 | 75 | – 5.2 | 90 | – 4.2 |
| | 0.2 | 65 | – 4.5 | 75 | – 3.5 |
| | 0.2 | 45 | – 3.1 | 55 | – 2.6 |
| | 0.2 | 30 | – 2.1 | 35 | – 1.6 |
| | 0.2 | 20 | – 1.4 | 25 | – 1.2 |
| | 0.2 | 10 | – 0.7 | 13 | – 0.6 |
| Hydrocarbon Gas Coolers | | | | | |
| | Temperature, °F | $U_b$ | $U_x$ | $U_b$ | $U_x$ |
| | 50 | 30 | – 2.1 | 35 | – 1.6 |
| | 100 | 35 | – 2.4 | 40 | – 1.9 |
| | 300 | 45 | – 3.1 | 55 | – 2.6 |
| | 500 | 55 | – 3.8 | 65 | – 3.0 |
| | 750 | 65 | – 4.5 | 75 | – 3.5 |
| | 1000 | 75 | – 5.2 | 90 | – 4.2 |
| Air and Flue-Gas Coolers | | | | | |
| Use one-half of value given for hydrocarbon gas coolers | | | | | |
| Steam Condensers | | | | | |
| (Atmospheric pressure & above) | | | | | |
| | | $U_b$ | $U_x$ | $U_b$ | $U_x$ |
| Pure steam (rf = .005) | | 125 | – 8.6 | 145 | – 6.8 |
| Steam with non-condensables | | 60 | – 4.1 | 70 | – 3.3 |
| HC Condensers | | | | | |
| | Pressure, psig | $U_b$ | $U_x$ | $U_b$ | $U_x$ |
| | 0° range | 85 | – 5.9 | 100 | – 4.7 |
| | 10° range | 80 | – 5.5 | 95 | – 4.4 |
| | 25° range | 75 | – 5.2 | 90 | – 4.2 |
| | 60° range | 65 | – 4.5 | 75 | – 3.5 |
| | 100° & over range | 60 | – 4.1 | 70 | – 3.3 |
| Other Condensers | | | | | |
| | | $U_b$ | $U_x$ | $U_b$ | $U_x$ |
| | Ammonia | 110 | – 7.6 | 130 | – 6.1 |
| | Freon 12 | 65 | – 4.5 | 75 | – 3.5 |

Note: $U_b$ is overall rate based on bare tube area and $U_x$ is overall rate based on extended surface.
Source: Gas Processor Suppliers Association, Engineering Data Book, 9th Edition.

**Figure 2.92** Typical heat-transfer coefficients for air coolers.

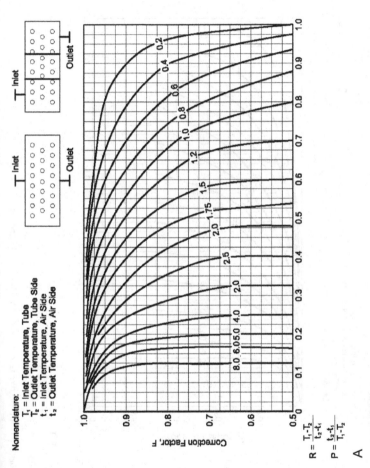

Figure 2.93 (A) LMTD correction factors; both fluids unmixed.

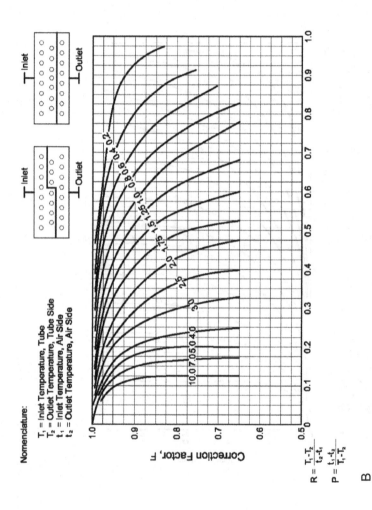

**Figure 2.93—**Cont'd (B) LMTD correction factor, two-pass cross-flow; both fluids unmixed.

**TABLE 2.8** External Area of Fin Tube per ft$^2$ of Bundle Surface Area (APSF) for 1-in. OD Tubes

| | ½-IN. HEIGHT BY 9 FINS/IN. | | 5/8-IN. HEIGHT BY 10 FINS/IN. | |
|---|---|---|---|---|
| Tube pitch | 2-in. △ | 2¼-in. △ | 2¼-in. △ | 2½-in. △ |
| Three rows | 68.4 | 60.6 | 89.1 | 80.4 |
| Four rows | 91.2 | 80.8 | 118.8 | 107.2 |
| Five rows | 114.0 | 101.0 | 148.5 | 134.0 |
| Six rows | 136.8 | 121.2 | 178.2 | 160.8 |

Control of cooled fluid temperature as the ambient air temperature changes

Usual mechanical considerations

Maintenance of fins

### Design Is a Trial-and-Error Calculation and Should Not Be Undertaken by a Novice

See also Figure 2.94.

▶ **COOLING TOWERS**

Specialized heat exchanger in which two fluids (air and water) are brought into direct contact with each other to affect the transfer of heat

### General Considerations

Entering air is cooled by being drawn through water with fans

Towers may have

Splash fill

Engineered/organized fill

No fill

Used where water is expensive to treat or in short supply

It is an aerial cooler preceded by an evaporating section (Figure 2.95)

**Figure 2.94** Air-cooled exchanger with finned tubes clogged with debris.

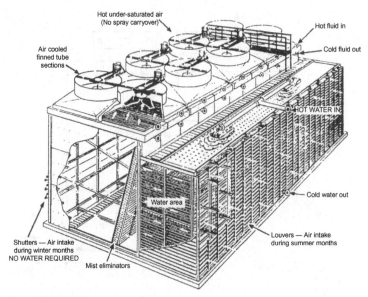

**Figure 2.95** Combination air–water cooler.

When air temperature is low enough the water may be shut off

Water rate may be decreased at intermediate temperatures

Higher initial cost but may offer a total cost savings in some applications

When equipped with controls to vary fan horsepower, the output of this unit offers flexibility at minimum operating cost

Water is cooled primarily through evaporation and humidifying the air

Figure 2.96 shows a "spray-filled" tower.

Cooling is accomplished by spraying a flowing mass of water into a rain-like pattern through which an upward moving mass flow of cool air is induced by the action of a fan

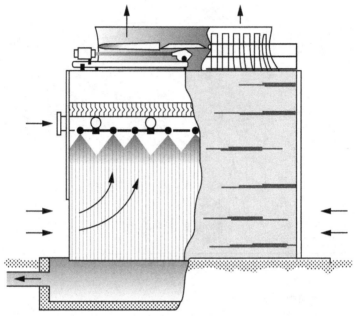

**Figure 2.96** Spray-filled cooling tower.

## ▶ OTHER TYPES OF HEAT EXCHANGERS
### Electric Heat Exchangers
See Figures 2.97–2.99.

### Heat Recovery Steam Generator
See Figures 2.100 and 2.101.

### Spiral Wound
See Figure 2.102.

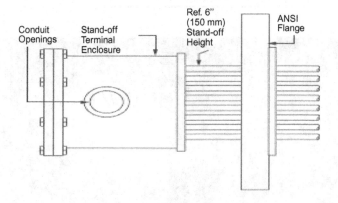

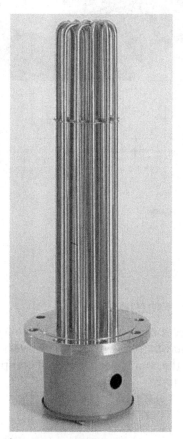

**Figure 2.97** Electric heat exchangers.

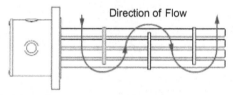

**Figure 2.98** Electric heat exchanger.

## Spiral
See Figure 2.103.

### ▶ HEAT EXCHANGER SELECTION

The proper selection of a heat exchanger involves many factors
Easy to choose one that will work; however, does it optimize cost of the total system without compromising operating reliability?

**Figure 2.99** Electric heat exchangers.

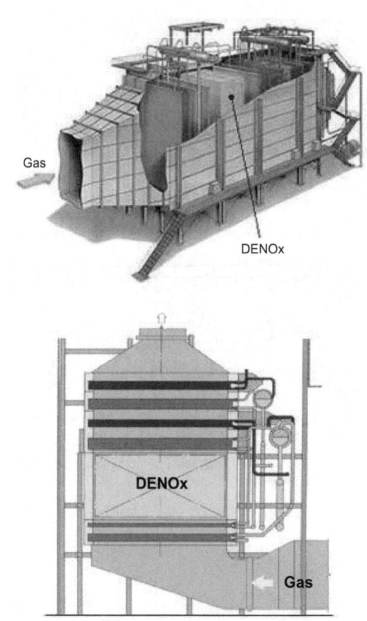

**Figure 2.100** Heat recovery steam generator.

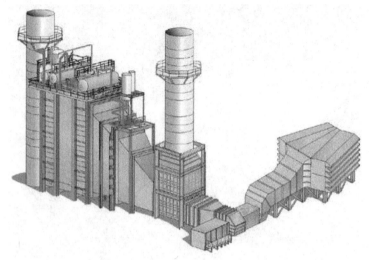

**Figure 2.101** Heat recovery steam generator.

## Guidelines We Should Follow

Do not specify or purchase a heat exchanger without considering its effect on the total process

Do not make the capital cost of the heat exchanger alone
the sole criterion for purchase

Acquaint the vendor with details of service and point out
that the choice will be made on both initial and operating
cost, not just initial capital cost alone

Use realistic pressure drop specifications, as this affects size
and cost. Allow as much pressure loss as economics dic-
tate for the actual system and not merely reproduce a
standard specification that might not apply

**Remember that the vendor knows his product but he knows
only as much about the application of his product as the
customer conveys to him**

**Figure 2.102** Spiral wound heat exchanger.

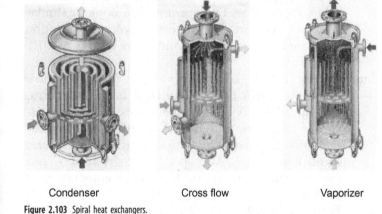

| Condenser | Cross flow | Vaporizer |

**Figure 2.103** Spiral heat exchangers.

▶ **EXAMPLE 3: LINE HEATER DESIGN**

**Given:**

Design a line heater for each of the 10 wells that make up the total 100 MMscfd field rate, that is, each well flows at 10 MMcfd

**Determine:**
1. Temperature for hydrate formation at 1000 psia
2. Heat duty for a single pass coil downstream of choke
3. Coil length for a 3-in. XX coil
   a. Calculate LMTD
   b. Calculate U
   c. Choose coil length
4. Fire tube area required and heater size (shell diameter, shell length, fire tube rating, coil length, and number of passes).

**Solution:**
1. Determine the temperature for hydrate formation at 1000 psia
   a. As shown in Table 2.9 below, using vapor equilibrium values from GPSA Engineering Data Book or similar reference we can determine the hydrate formation temperature at 1,000 psia and temperatures of 50°F and 70°F.

Interpolating linearly, $\Sigma$ ($Y/K_{V\text{-}S}$) = 1.0 at 66.9°F
   b. Using the GPSA Engineering Data book, or equivalent, specific gravity vs temperature curves at a pressure of 1,000 psia and specific gravities of 0.6 and 0.7 one can interpolate a hydrate formation temperatures of 60°F and 64°F. At a specific gravity of 0.67 one can interpolate a hydrate formation temperature of 62.8°F.
   From Table 2.9, S = 0.67
   At S = 0.6, P = 1000 psia, hydrate temperature = 60°F
   At S = 0.7, P = 1000 psia, hydrate temperature = 64°F
   By interpolation, hydrate temperature at S = 0.67 is 62.8°F
2. Determine the process heat duty

The temperature at the outlet of the heater should be between 5 and 15°F above the hydrate formation temperature.

**TABLE 2.9** Hydrate Formation Temperature Determination

| | Y | VALUES AT 1000 PSIA | |
|---|---|---|---|
| COMPONENT | MOLE FRACTION | 50°F | 70°F |
| $N_2$ | 0.0144 | — | — |
| $CO_2$ | 0.0403 | 0.60 | — |
| $H_2S$ | 0.000019 | 0.07 | 0.38 |
| $C_1$ | 0.8555 | 1.04 | 1.26 |
| $C_2$ | 0.0574 | 0.145 | 1.25 |
| $C_3$ | 0.0179 | 0.03 | 0.70 |
| $iC_4$ | 0.0041 | 0.013 | 0.21 |
| $nC_4$ | 0.0041 | 0.145 | 1.25 |
| $iC_5$ | 0.0020 | — | — |
| $nC_5$ | 0.0013 | — | — |
| $C_6$ | 0.0015 | — | — |
| $C_{7+}$ | 0.0015 | — | — |
| | 1.0000 | $\Sigma \, (Y/K_{v\text{-}s}) =$ 2.226 | $\Sigma \, (Y/K_{v\text{-}s}) =$ 0.773 |

With a hydrate formation temperature of 62.8°F, select an outlet temperature of 75°F.

a. Temperature drop through choke

    Flowing tubing pressure, PSIG    4000

    Heater inlet pressure, PSIG       1000

    $\Delta P$ through choke, PSIG         3000

    $\Delta T = 79°F$ from Figure. Curve based on 20 bbl/MMscf condensate. Well has 60 bbl/MMscf, therefore $\Delta T_{corrected} = 79 - 20 = 59°F$

    Accordingly, the heater inlet temperature $= 120 - 59 = 60°F$

b. Gas duty

    Flowing pressure, P, psia        1015

    $P_c$, psia (from Table)          680

    Reduced pressure, $P_R = P/P_C$    1.49

    Heater inlet temp, °F           61

    Heater outlet temp, °F         75

    Average temp, °F $(61 + 75)/2$    68

$$\text{Average temp, } °R \qquad\qquad 528$$

$$T_c, °R \text{ (Table)} \qquad\qquad 375$$

$$T_R = T/T_c \qquad\qquad\qquad 1.41$$

$$q_g = 41.7(\Delta T)C_g Q_g$$

where

$q_q$     =     gas heat duty

$\Delta t$     =     $t_{out} - t_{in}$

Because flow through coil is a constant pressure process, we have:

$\Delta t$     =     $t_{out} - t_{in} = 75 - 61$    =    $14°F$

$Q_g$     =     gas flow, MMscfd     =    10

$C_g$     =     gas heat capacity, Btu/Mscf-°F

Calculation of $C_g$

$C_g$     =     $2.64 (29)(C) + (\Delta C_P)$

where

$C$     =     gas-specific heat, Btu/lb °F

From Figure 1.32, C at 68°F is = 0.50

$\Delta C_P$, from Figure 1.33 (at $T_R = 1.41$ and $P_R = 1.49$) = 2.6

$S$     =     0.67

$C_g$     =     $2.64(29)(0.67)(0.50) + 2.6)$

       = 32.51 Btu/Mscf-°F

$g_g$     =     $(41.7)(14)(32.51)(10)$

       = 190 Mbtu/hr

c. Oil duty

$q_0$     =     $14.6 (SG)(T_2 - T_1)C_o Q_o$

where

$SG$     =     0.77

$Q_o$     =     oil flow rate, BPD

       =     (60 BBL/MMSCF)(10 MMscfd)

$t_2 - t_1$     =     $75 - 61$

       =     $14°F$

$C_o$     =     oil-specified heat, Btu/lb °F

From Figure 1.31 at 68°F (52.3° API)

$C_o$     =     0.48

$q_o$     =     $14.6 (0.77)(14)(0.48)(600)/100$

       =     46 Mbtu/hr

d. Water duty

$q_w = 14.6\,(T_2 - T_1)Q$

Gas is saturated with water at 8000 psig (shut-in BMP) and 224°F (BHT). From McKetta Wehe graph, we have

LB water/MMscf of wet gas at reservoir conditions (8000 psig and 224°F)                     260

LB water/MMscf of wet gas at 1000 PSIG and 75°                                                           28

Water to be heated                                                                                                         232

Water quantity $= Q_w$

$$= \left(232\,\frac{lb}{MMscf}\right)\left(\frac{10MMscf}{62.4 lb/ft^3}\right)\left(\frac{7.4gal}{ft^3}\right)\left(\frac{BBL}{42gal}\right)$$

$$= 6.6 \text{ BPD}$$

$Q_w = (14.6)\,(75 - 61)(6.6)$

$\quad\;\; = 1.3 \text{ MBtu/hr}$

e. Total process duty

$q = q_g + q_0 + q_w$

$\;\;\, = 90 + 46 + 1.3$

$\;\;\, = 237 \text{ MBtu/hr}$

3. Calculation of coil length

  a. Calculation LMTD

Temperature of bath is 190°F

$$\Delta t_1 = 190 - 61 = 129$$
$$\Delta t_2 = 190 - 75 = 115$$
$$\text{LMTD} = \frac{14}{\log_e\left(\frac{129}{115}\right)}$$
$$= 122°F$$

b.  Calculate U

$$\frac{1}{u} = \frac{1}{h_0} + R_0 + \frac{L}{K} + R_i + \frac{A_0}{h_{iA_1}}$$

Uses $R_o + R_1 = 0.003$ hr ft$^2$°F/Btu

$$h_0 + 116 \left[ \frac{k^3 C^2 \rho \beta \Delta t}{\mu d_0} \right]^{0.25}$$

where

K   =  0.39 (Figure 1.26)
$C_p$  =  1 Btu/lb°F (Figure 1.24, 1.31, 1.33)
$\rho$   =  60.35 lb/ft$^3$ (Table, 1/0.01657)
$\beta$   =  0.0024 1/°F(Table 1.5)
$\mu$   =  0.32 cp (Figure)
$\Delta t$  =  122°F
$d_o$   =  3.5 in

$$h_0 = 116 \left[ \frac{(0.39)^3 (1)^2 (60.35)(0.0024)(122)}{(0.32)(3.5)} \right]^{0.25}$$

$$= 114 \text{ Btu/hr ft}^2 \text{ P}$$

For 3-in. XX pipe A-106B
    L   =   0.60 in.–0.50 ft (Table)
    K   =   30 Btu/hr ft °F (Table)
        =   (for 90 CU and 10 Nl)
    $A_0$  =   0.916 ft$^2$/ft (Table)
    $A_i$  =   0.602 ft$^2$/ft (Table)

$$h_i = \frac{0.022K}{D}\left(\frac{DG}{\mu_e^{0.8}}\right)\left(\frac{C_{\mu e}}{K^{0.4}}\right)\left(\frac{\mu_e}{\mu_{ew}}\right)^{0.16}$$

where

| | | |
|---|---|---|
| D | = | 2.30 in = 1.92 ft (Table) |
| A | = | $\pi D^2/4 = 0.0289$ ft$^2$ |
| K | = | 0.017 Btu/hr ft°F 1.25 (Figure) |
| | = | 0.021 Btu/hr ft°F |

$$Gas\ Flow = (10MMscf)\left(\frac{D}{24hr}\right)\left(\frac{lb/mol}{379scf}\right)\left(\frac{19.4lb}{lb/mol}\right)$$

$$= 21,238\ lb/hr$$

$$Oil\ Flow = \left(\frac{608BBL}{D}\right)\left(\frac{D}{24hr}\right)\left(\frac{350lb}{BBL}\right)(0.77)$$

$$= 6827\ lb/hr$$

$$Water\ Flow = \left(\frac{6.6BBL}{D}\right)\left(\frac{D}{24hr}\right)\left(\frac{350lb}{BBL}\right)$$

$$= 96\ lb/hr$$

$$G = \text{mass velocity of fluid}$$

$$= \frac{(21,328) + (6,827) + (96)}{0.0299}$$

$$= 977,560\ lb/hr - ft_2$$

$$c = \left(\frac{32.51btu}{Mscf°F}\right)\left(\frac{6,827\ lb}{28,252lb}\right) + \left(\frac{1BTU}{lb°F}\right)\left(\frac{96lb}{28,252lb}\right)$$

$$+ \left(\frac{0.48btu}{lb°F}\right)\left(\frac{6,827lb}{28,252lb}\right) + \left(\frac{1btu}{lb°F}\right)\left(\frac{96lb}{28,252lb}\right)$$

$$= 0.60\ Btu/lb°F$$

$$\mu \quad = \quad 0.0134 \text{ cp (at } 60°F)(2.4) \quad = \quad 0.0322 \text{ lb/hr ft}$$
$$\mu_W \quad = \quad 0.0142 \text{ cp (at } 129°F)(2.4) \quad = \quad 0.0341 \text{ lb/hr ft}$$

$$0h_i \;=\; \frac{(0.022)(0.021)}{0.192}\left(\frac{(0.192)(977.560)}{0.0322}\right)^{0.8}$$
$$+\left(\frac{(0.6)(0.0322)}{0.021}\right)^{0.4}\left(\frac{0.0322}{0.034}\right)^{0.16}$$
$$=\; 595 \text{ Btu/hr ft}^{2°}F$$

$$\frac{1}{u} = \frac{1}{114} + 0.003 + \left(\frac{0.05}{30}\right) + \left(\frac{0.9162}{(595)(0.6021)}\right)$$

U      =    62.5 Btu/hr ft °F
Estimated U from Figure
   U    =    106 Btu/hr ft °F
Use U   =    62.5 Btu/hr ft² °F
c.  Calculate coil length

$$L = \frac{12q}{\pi(LMTD)Ud}$$

where

q    =    237 MBtu/hr (total process heat duty)
L    =    122°F
U    =    62.5 Btu/hr ft²°F
d    =    3.5 in.

$$L \;=\; \frac{(12)(237,000)}{\pi(122)(62.5)(3.5)}$$
$$=\; 39.9 \text{ ft}$$

4. Calculate fire-tube area required
   For heat transfer to water use 10,000 Btu/hr ft$^2$
   Flux rate

$$A = \frac{237,000btu/hr}{10,000btu/hr - ft^2}$$
$$= 23.7 \text{ ft}^2$$

Estimate shell size
   Assuming a 10′-0″ shell, then four passes of 3″XXH are required. This will require a 30-in. OD shell for the coils and fire tube.

5. Summary of line heater size

| | |
|---|---|
| Heater duty | 250 MBtu/hr |
| Coil size | 3″XXH |
| Minimum coil length | 33.9 ft |
| Minimum fire-tube area | 23.7 ft |
| Shell size | 30″ OD × 10′-0″ F/F |

▶ **EXAMPLE 4: SHELL-AND-TUBE HEAT EXCHANGER DESIGN**

**Given:**

| | |
|---|---|
| Inlet | 100 MMscfd at 0.67 SG (Table) |
| | $T_1 = 175°F$ |
| | $P_1 = 1000$ psig |
| | Water vapor in gas = 60 lb/MMscf |
| | 15 bbl water/MMscf |
| Outlet | $T_2 = 100°F$ |
| | $P_2 = 990$ psig |
| | Water vapor in gas = 28 lb/MMscf |
| Seawater | $T_3 = 75°F$ |
| | Limit temperature rise to 10°F |

Use 1-in. OD 10 BWG tubes on 1¼-in. pitch

Design a seawater cooler to cool the total stream from the example field in its later stages of life from a flowing temperature of 175°F to a temperature of 100°F to allow further treating

**Determine:**
Calculate water vapor condensate
Calculate heat duty
Determine seawater circulation rate
Pick a type of exchanger and number of tubes
Check to assure water velocity in tubes does not exceed 15 ft/s

**Procedure:**
Calculate heat duty
Determine fluid in shell/tube
Assume/calculate overall heat transfer coefficient
Choose
  (a) number of shell passes
  (b) number of tube passes
Correct LMTD
Select tube diameter and length
Calculate the number of tubes
Determine shell diameter
Determine type of exchanger
Determine tube velocity

**Solution:**
1. Calculate free water and water vapor flow rates
   Water flow rate inlet
   Free water = (100 MMscfd)(15 bbl/MMscfd)
             = 1500 bwpd

Water vapor condensed

$$= \left( \frac{(60 - 28)lb}{MMscf} \right) \left( \frac{100 MMscf}{d} \right)$$

$$= 3200 \text{ lb/day}$$

$$Q_w = (3200 lb/d) \left( \frac{bbl}{350 lb} \right)$$

$$= 9 \text{ bwpd}$$

Water flow rate outlet

| | |
|---:|---|
| 9 | bwpd |
| 1500 | bwpd |
| 1509 | bwpd |

2. Calculate heat duty
   a. Gas duty

$T_1$ = 635°R(175°F inlet)

$T_2$ = 560°R(100°F outlet)

$T_{AV}$ = 597.5°R

$P_C$ = 680 psia (Table)

$P_R$ = P/$P_C$ = 1.47(1000/680)

$T_C$ = 370°R (Table)

$T_R$ = $T_{AVG}$/$T_C$ = 1.62(597.5/370)

$q_g$ = 41.7 $\Delta T C_g Q_g$ (sensible heat of natural gas)

$C_g$ = 2.64(29 SG + $\Delta C_P$)

C = 0.528 Btu/lb °F (Figure 2-16, Textbook)

$\Delta C_P$ = 1.6 Btu/lb-mol °F (Figure)($P_R$ & $T_R$)

S = 0.67 (Table)

$C_g$ = 2.64 [(29)(0.67)(0.528) + 1.6]

= 31.3

$q_g$ = 41.7(100-175)(31.3)(100)

= 9,789,000 Btu/hr

b. Condensate duty

$q_O$ =    14.6 (S.G.) $\Delta T C_O Q_O$

$C_O$ =    0.535 Btu/lb°F (Figure)

$q_O$ =    14.6 (0.77) (100-175)(0.535)(6000)

=    $-2,707,000$ Btu/hr

c. Free water duty

$q_w$    =    14.6 $\Delta T Q_w$

=    (14.6)(100-175)(1509)

=    1,652,000 Btu/hr

d. Water latent heat duty

$q_{lh}$    =    (3200 lb/D)(1/24)

=    133 lb/hr

$\lambda$    =    $-996.3$ Btu/lb (Table, 170°F)

$q_{lh}$    =    (133) ($-996.3$)

=    $-133,000$ Btu/hr

e. Total heat duty

$q$    =    $q_g + q_0 + q_w + q_{lh}$

=    $-9,789,000-2,707,000-1,652,000-133,000$

=    $-14,281,000$ Btu/hr

3. Water circulation rate

$$q_w = 14.6 \, (t_2 + t_1) Q_w$$

Rearranging and solving for $Q_w$

$$Q_w = \frac{q_w}{14.6(t_2 - t_1)}$$

Limit $\Delta T$ for water to 10°F to limit scale

$Q_w$    =    $(14.3 \times 10^6)(10)$

=    97,945 bwpd

=    2858 gpm

4. Heat exchanger type and number of tubes

Choose TEMA R because of large size (offshore)

Select type AFL because of low temperature change and LMTD correction factor

- A - channel and removal cover (used commonly with fixed tube sheet)
- F - two-pass shell with longitudinal baffle
- L - fixed tube sheet
  - least expensive
  - fewer gaskets
  - individual tubes can be replaced
  - used in clean shell fluids and low differential temperatures ($<200°F$)
  - shell cannot be cleaned and inspected
  - bundle cannot be replaced

If brackish water, use higher nickel content for tube material, therefore, 70/30 $C_U N_I$

Because the water is corrosive and may deposit solids, flow water through tubes

**U = 90 Btu/hr °F (Table)(water with 1000 PSI gas)**

**Calculate LMTD**

$$t_1 = 175 \qquad t_2 = 100$$
$$t_3 = 75 \qquad t_4 = 85$$
$$\Delta t_1 = 175 - 85$$
$$= 90$$
$$\Delta t_2 = 100 - 75$$
$$= 25$$

$$LMTD = \left[ \frac{90 - 25}{\log_e \left( \frac{90}{25} \right)} \right]$$
$$= 50.7°F$$

Correction factor (from Figure)

$$P = \frac{85 - 75}{175 - 75} \qquad R = \frac{175 - 100}{85 - 75}$$

$$= 0.1 \qquad\qquad = 7.5$$

$$\begin{aligned}
\text{F} \quad &= \quad 0.95 \\
\text{MTD} \quad &= \quad (50.7)(0.95) \\
&= \quad 48.2°\text{F}
\end{aligned}$$

Calculate number of tubes

$$N = \frac{q}{UA'(LMTD)L}$$

Assume

$L = 40$ ft

$A = 0.2618$ ft$^2$/ft (Table)

$$N = \frac{14.3 \times 10^6}{(90)(0.2618)(48.2)(40)}$$

$= 315$ tubes

From Table for 1-in. OD, 1¼-in. SQ pitch, fixed tube sheet, two passes, shell ID = 29 in.

If, Assume    L    =    20 ft (instead of 40 ft)

Then          N    =    629 tubes

Shell ID           =    39 inches

Use 39-in. ID × 20 ft long with 682-in. OD, 10 BWG tubes 1¼-in., square pitch with two tube passes. Size 39-240 type AFL

5. Check water velocity

$$V = 0.012 \frac{Q_w}{d^2}$$

There are four passes. Thus 682/2 tubes are used in each pass

$$Q_w per\ tube = \frac{(4)(97,945)}{682}$$

$$= 227\ bwpd$$
$$D = 0.732\ in.\ (from\ Table)$$
$$V = \frac{(0.012)(574)}{(0.732)^2}$$

$$= 6.4\ ft/s,\ which\ is\ less\ than\ 15\ ft/s$$

### ▶ COMMENTS ON EXAMPLE 4

Once the heat exchanger has been chosen and the flow per tube, a more precise "U" can be determined

More than 30% of the heat duty was required to cool the water and condensate

If liquids had first been separated
  Less heat duty required
  Smaller heat exchanger
  Lower seawater flow rate
  Therefore, cooler usually placed downstream of first separator
  Often an aerial cooler is used in this service

Outlet temperature of 100°F may be sufficient if only compression and dehydration are required

## If Gas Treating Required ($H_2S$ or $CO_2$)

Amine unit will heat gas 10–20°F

Thus, gas dehydration would be difficult at elevated temperature

Therefore, better to cool gas lower initially so that still below 110°F at glycol dehydration

Often not possible

Cooling water not available

Ambient air conditions are in the 95–100°F range

May require aerial cooler to cool gas before treating and before dehydration

▶ **IN-CLASS EXERCISES**

1. Match the following heat transfer equations with their description.
    a. Convection ___                        $q = UA\Delta T$
    b. Conduction ___                        $q = hA\Delta T$
    c. Radiation ___                          $q = \Delta T/\sum R$
    d. Overall heat transfer equation ___   $q = A\sigma T^4$
    e. Thermal resistance ___               $q = (kA/L)\,\Delta T$

2. What effect do fouling factors ($r_0$ and $r_i$) have on the overall heat transfer coefficient?

    _____

    _____

    _____

3. What is the advantage of high velocity in heat exchangers?

    _____

    _____

4. What are the three assumptions used in the derivation or LMTD?
   a. _____
   b. _____
   c. _____

5. Double-pipe exchangers are best suited for application involving large heat transfer surfaces. True or False

6. What is the main difference between floating head and fixed tube sheet exchangers?_____
   _____

7. What is the different between tube pitch and tube clearance?
   _____
   _____

8. What are the purposes of baffles in shell-and-tube exchangers?_____
   _____

9. TEMA Class "R" heat exchangers are more heavy duty than Class "C" heat exchangers. True or False

10. What is the purpose of fins in air-cooled exchangers?
    _____
    _____

11. Compare the fuel efficiency of a fired heater using only a radiant section with that of a fired heater using a radiant and convection section.
    _____
    _____

12. Indicate which type of heat transfer equipment is described: double pipe, shell and tube, plate fin, plate and frame, air cooler, fired heater, or waste heater
    _____Finned tube-and-fan system
    _____Pipe within a pipe
    _____Radiant section
    _____Most are equipped with baffles

_____Bundle of tubes within a shell
_____Clean services only
_____Convection section only
_____Shield tubes
_____True countercurrent
_____Very close approaches
_____Efficiency is dependent on fan performance
_____Uses heater or turbine exhaust

13. Given:

Inlet condition

| | |
|---|---|
| $T_1$ | $= 165°F$ |
| $P_1$ | $= 600$ psig |
| Moisture content | $= 70$ lb $H_2O$/MMSCF |
| $Q_w$ | $= 20$ BBL/MMSCF |

Outlet conditions

| | |
|---|---|
| $T_2$ | $= 100°F$ |
| $P_2$ | $= 585$ psig (minimum) |
| Moisture content | $= 25$ lb $H_2O$/MMSCF |

Seawater

| | |
|---|---|
| $T_3$ | $= 72°F$ |

Determine:

Design a shell-and-tube heat exchanger that will cool 175 MMSCFD (S = 0.65) of gas from a flowing temperature of 165° to 100°F

Assume 1-in. OD 10 BWG tubes on 1¼-in. Δ pitch are available and the process is located offshore

14. What type of heat exchanger is the workhorse for separation, heating, and cooling?

15. What is the difference, if any, between cocurrent (parallel-flow) heat exchangers and countercurrent flow heat exchangers?

16. Heat exchangers are not susceptible to problems because they have no moving parts. True or False

17. What has the Tubular Exchanger Manufacturers Association developed for describing shell-and-tube heat exchangers?
18. For shell-and-tube heat exchangers, what is the typical shell type choice? Why?
19. Explain how a cooling tower, which is a specialized heat exchanger, transfers heat.
20. Plate-and-frame exchangers are used mostly in upstream installations. True or False
21. What TEMA size would a 40-ft straight length U-tube bundle, 6-ft shell diameter, with a single shell pass and a removable shell cover be?
22. What is the most important distinguishing feature describing heat exchangers?
23. During exchanger design, what is the setting of the tube count per pass based on?
24. Why is it difficult to clean heavy deposits from the inside on U tubes?
25. Compact heat exchangers get their name from their small size. True or False?

# Tubular Heat Exchanger Inspection, Maintenance, and Repair

**3**

## ► GENERAL INFORMATION

### Introduction

Abbreviations of organizations and publications used throughout this section

| | |
|---|---|
| American National Standards Institute | ANSI |
| American Petroleum Institute | API |
| American Petroleum Institute, *Pressure Vessel Inspection Code* | API 510 |
| American Petroleum Institute, *Inspection of Pressure Vessels, Recommended Practice 572* | API 572 |
| American Petroleum Institute, *Material Verification Program for New and Existing Alloy Piping Systems* | API 578 |
| American Petroleum Institute, *Damage Mechanisms* | API 571 |
| American Society of Mechanical Engineers | ASME |
| ASME, *Boiler and Pressure Vessel Code* | ASME Code |
| American Society for Nondestructive Testing | ASNT |
| American Society of Testing Materials | ASTM |
| American Welding Society | AWS |
| Heat Exchange Institute of Cleveland, Ohio | HEI |
| Heat Exchange Institute, *Standards for Closed Feed-water Exchangers* | HEI CFHS |

| Heat Exchange Institute, *Standard for Power Plant Heat Exchangers* | HEI, PPS |
| National Board of Boiler and Pressure Vessel Inspectors    National Board National Board Inspection Code | National Board Code |
| Standards of the Tubular Exchanger Manufacturers Association | TEMA |

## Inspector Qualifications

Few heat exchanger users can justify maintaining a staff inspector qualified only to inspect tubular heat-transfer equipment

Desirable attributes of inspectors responsible for inspecting tubulars

Experienced in heat exchanger construction

Familiar with exchanger fabrication and repair-shop practices

Trained in reading and understanding heat exchanger construction drawings

Knowledgeable about the applicable pressure safety and inspection codes, API Inspection Code sections, and TEMA Standards

Familiar with and skilled in using inspection tools and instruments

Trained to make, record, and interpret measurements

Experienced with, and knowledgeable about, destructive and nondestructive testing and examination

Knowledgeable about materials used in constructing heat exchangers and applicable materials specifications

Familiar with erosion, corrosion, erosion–corrosion, and evidence failure and deterioration of heat exchanger parts

Trained and qualified by an inspection organization, e.g., TUV or the National Board, in pressure-vessel inspection

Certified or commissioned as a qualified inspector by a private or public agency in accordance with national, international, or industry standards

Aware of hazards when inspecting in operating facilities and in heat exchanger factories

Conscious of safety requirements when performing inspections

Aware that inspections and inspection records are essential to proper maintenance, repair, and life extension of tubular heat exchangers

### Roles of Code Authorized Inspectors and Noncode Inspectors

Responsibilities of authorized inspectors

Bound by the responsibilities delineated in the applicable pressure safety construction and inspection codes

Primary concern is the safety of pressure-containing parts of the exchanger and the welds of any nonpressure-bearing parts to the pressure envelope

Responsibilities of authorized inspectors, listed in the ASME Code and the NBIC, are typical of most pressure-safety codes. Some of these are:

Verifying that design calculations have been made in accordance with the code rules or that the user's and manufacturer's design reports are on hand and are being followed

Making sure that all pressure parts, parts welded to them, and welding materials conform with the prevailing pressure vessel code requirements

Confirming that only qualified personnel perform welding and that the welding conforms with qualified procedures

Ascertaining that all manufacturing, repair, or alteration methods are in accordance with the pressure vessel code or inspection code rules

Determining that all nondestructive examinations are performed and interpreted by qualified examiners and accepting or rejecting work examined in accordance with code criteria

Verify that the manufacturer follows the quality assurance and quality control manuals submitted to the certifying authority

Inspect fit-ups and pressure welds

Witness specified leak and pressure tests required by the code

Sign the appropriate report form to indicate acceptance of the newly constructed pressure vessel. In the United States and Canada, this is the ASME Code Manufacturer's Data Report.

Sign the appropriate report form to indicate acceptance of a repaired or altered pressure vessel. In the jurisdiction where the National Board Code applies, this is one of the "R" report forms for alterations or repairs

Verify that replacement exchangers are so constructed that they perform as well as the originals

Authorized inspectors do not

Concern themselves with any specification or manufacturer's standard to which the pressure safety or inspection code does not refer

Pay attention to deviations from approved drawings that are not code violations, for example

Cocked or slightly misplaced nozzles

Misaligned support bolt holes

Under- or oversized baffles and supports

Verify that a repair or alteration will not affect an exchanger's operation adversely

Determine that an exchanger is well constructed for its performance requirements

Ascertain whether the clearance between replacement bundles and existing shells is smaller or larger than in the original exchanger. They may not consider that if the clearance is

- Too small, the bundle may be damaged during insertion
- Too large, it will permit more-than-designed quantities of shell fluid to flow through the space between the shell and the baffles, thereby reducing performance

Responsibilities of noncode inspectors

Responsibilities partially overlap those of code authorized inspectors

Primary duties are to

Verify that the work meets the purchaser's specifications

Make sure that all dimensions and fits of parts conform with agreed-on specifications, standards, and approved drawings

Make sure that the manufacturer stores and handles materials and parts in ways that will not shorten exchanger life

Be present at critical times, such as tubing the cage and stabbing the bundle into the shell

Deal with inadvertent factory errors

Expedite materials procurement and production (sometimes)

Hold points and coordination with the shop

User or purchaser's inspector must

- Coordinate with the manufacturer the hold points required by the purchase order and inspection brief
- Work with the manufacturer to keep track of the approach of hold points and to avoid delaying work

Shops must
- Stop construction at hold points that the authorized inspector designates
- Arrange their schedules to coordinate with the authorized inspector's schedule

### Preinspection Conferences

Prepare inspectors and make sure that they understand the guidelines embodied in the inspection brief

Subjects depend on what is to be inspected:

New equipment or replacement parts

Mothballed or used units

Exchangers undergoing scheduled or unplanned outages

Aim for new exchanger and replacement parts to ensure that all concerned individuals are fully aware of and understand the brief

Each participant should understand

What is to be inspected

At what stage of production it is at

Acceptance and rejection criteria

Also used to

Review the procedure to be followed for
- Final documentation
- Release for shipment
- Special shipping requirements (e.g., monitoring impacts during transit)

Aim for mothballed or used tubulars to make sure that the participants know the considerations for

Restoring a unit to service

Putting a mothballed one into a new service

Procuring a used exchanger as a replacement

### Inspection Tools and Instruments

Tools and instruments used for inspecting tubulars and replacement parts under construction are

The same as those used for inspecting operating units or
those disused units being considered for service

Used to make sure dimensions, thicknesses, locations,
construction, etc., conform to the approved drawings
and specifications and standards

Maintenance inspecting during outages or surveying mothballed or used equipment requires more elaborate tools
and methods to assess the condition of the parts and to
measure thickness

Aim is to try to determine the present condition of previously accepted equipment that has deteriorated because
of its service or exposure to the elements

Table 3.1 suggests parts to be inspected and tools and instruments that should be available to the inspector

S    Apply mostly to shop inspections

F    Apply mostly to field inspections

SF   Apply to both shop and field inspections

Table 3.2 is a list of tools that should be kept in well-
organized toolboxes for inspecting in the field

Table 3.3 lists equipment and instruments used primarily
for maintenance inspections

**Inspection Reports and Data Collection**

Inspector's reports and data collected should be made part
of the permanent maintenance file

Knowing what took place during the work is invaluable for
troubleshooting causes of failure in the future

The following items should be included in the inspector's
report

An as-built dimension drawing showing approved dimensions marked to show all deviations

A weld seam location sketch

A location sketch of tubes plugged by the manufacturer or
repair concern, with supporting information about the
type of plug used and the plugging procedure. Should

**TABLE 3.1** Parts to Examine vs. Tools and Instruments to Use

| | THINGS TO EXAMINE | WHAT TO EXAMINE IT WITH |
|---|---|---|
| S | Baffle cuts | Scale |
| SF | Bolt elongation | Micrometer calipers: sonic micrometer |
| S | Bolt-hole orientation | Bolts used as pins in bolt holes; level, square, and straightedge |
| SF | Channel cover flatness | Straightedge with mounted dial indicator |
| SF | Tube sheet flatness | |
| SF | Flange flatness | |
| S | Surface finishes | Tactile comparison samples; tactile comparison gage; optical comparator |
| SF | Flange scalloping (distortion between bolt holes) | Feeler gage for assembled flanges; straightedge with mounted dial indicator for disassembled flanges |
| SF | Tube sheet ligament distortion | Internal dial calipers; strain gages |
| S | Nozzle and support angularity | Level, straightedge, or square |
| F | Pass-partition bending | Scale, straightedge with mounted dial |
| SF | Tube bowing | High-intensity light |
| F | Tube condition | Borescope with video recorder; eddy current test system; magnetic resonance test system; pulsed ultrasonic beam inspection system; internal rotary inspection scan (IRIS) |
| SF | Tube diameter, ovality, and thickness | Hole gage; internal dial calipers; outside micrometer calipers |
| S | Tube holes in tube sheets | Go/no-go gage |
| SF | Tube-to-tube sheet joints | 10-power magnifying glass; 30-power illuminated pocket microscope; bubble test kit; fluid penetrant examination kit; halogen leak sniffer (halogen diode tester) and accessories; helium leak sniffer (mass spectrometer) |
| S | Tube sheet-to-tube sheet parallelity | Steel measuring tape; strung piano wire; transit; split-image transit |
| S | Tube sheet rotation about longitudinal axis | Level, straightedge, plumb line, and protractor |
| S | Tube sheet waviness | Straightedge with mounted dial indicator |
| F | Tube-wall thickness | Before installation—micrometer calipers; after installation—IRIS, eddy current tester |
| SF | Shell interior condition | Borescope and video recorder; scrapers; pit gage |
| F | Shell expansion joint angular deflection and squirm | Trammels and scale, level and straight-edge |

**TABLE 3.2** Toolbox Inspection Equipment for Shop Inspections and In-Service Equipment Inspections

Sulfur-free marking crayon
Low-voltage-powered drop light with extension (explosion proof where required)
Feeler gage set
Flashlight (explosion proof where required)
Inspector's hammer (spark proof where required)
Lightweight ball peen hammer (spark proof where required)
Hook gage to measure internal welds and cavities
10-power magnifying lens; 30-power illuminated pocket microscope
Inspection mirrors (jointed, magnifying, etc.)
Pit depth gage to determine depth of pit corrosion
Scrapers: steel for scraping unmachined surfaces; brass or copper for scraping machined surfaces
Wire brushes (use nonferrous or stainless where required)

**TABLE 3.3** Larger Supplementary Inspection Equipment (Not Carried in Toolbox)

Borescope set
Borescope video recorder, keyboard, and console
Eddy current testing system equipment
Fluid penetrant examination kit and special lamps
Halogen leak sniffer and accessories
Helium leak sniffer and accessories
Hydroblasting equipment
Magnetic particle testing equipment
Pressure-testing pumps, gages, and recorders
Trepanning, hole-drilling and hole-plugging equipment
Telescopic transit
Split-image transit
Ultrasonic inspection apparatus (IRIS and pulsed types)
X-ray or isotope camera, shielding, radiation-zone exclusions, warnings, personnel radiation recorders and badges (dosimeters)
Video camera (hand held)

include sketches and descriptions of how broken tubes are stabilized in a bundle

A tube heat map when more than one heat of tubing is used. In addition to showing the location of the different heats of tubing, it should show a record of torque values or hydro-expanding pressures used for expanding the tubes in each heat

Certified copies of mill test reports and whatever documents the shop has used to identify where the material has been used. The inspector should verify that the heat numbers are stamped or stenciled on all critical parts and correspond with the mill test reports. Mill test reports should be examined for conformity with applicable specifications.

Rubbings of the nameplate, whether a manufacturer's or a repair or alteration one, or a legible photograph of the nameplate

For units constructed to ASME Code, a signed original of the U-1, U-2, A-1, and A-2 Manufacturer's Data Report forms and attached partial data reports, if any, or the NBIC form "R" Repair or Alternate Certificate. For exchangers constructed to other pressure vessel safety codes, equivalent documentation.

Records of nondestructive examinations

Records and reports of leak and hydrostatic tests

Manufacturer's Design Report (for ASME Code requiring such reports); copies of ASME Code calculations for exchangers constructed to Section VIII, Division 1, calculations required by and other national pressure safety codes that apply

### Inspection Records as a Maintenance Tool

Collected records of new equipment and parts inspection, inspection online, and during outages are essential for troubleshooting causes of failures and for planning repairs, alterations, and replacements

The following example shows how an inspection record can be used to establish the cause of a failure

A heater was installed with 18-8 austenitic stainless steel, welded tube sheets

To make sure that no fluid leaked from the shell side into the tubes, at regular intervals tube-to-tube sheet joints were bubble tested with 50 psi (345 Pa) nitrogen in the shell and commercial, chloride-free bubble former was spread on the tube sheet faces

Shortly after one such test circular transverse cracks were observed in the tubes where they penetrated the rear face of the tube sheet and for a short distance beyond that point

An examination of the report revealed the comment that "soap flakes and tap water" were used as the commercial bubble former wasn't available

Metallurgical analysis disclosed that the failures were the result of stress-corrosion cracking (SSC)

The following three conditions must be met for SCC to occur:

A material sensitive to SCC

A level of tensile stress in the material

An environment that promotes SCC

Austenitic 18-8 alloys are sensitive to SCC

The level of tensile stress in the region of the transition from expanded to unexpanded tube is in the order of 90% of the yield stress

The environment was provided by the high chloride ion content of the tap water and soap

Without an examination of the record, all that could have been found by metallurgical analysis would have been SCC and the presence of high concentration of chlorides

## ► ASIAN, EUROPEAN, AND NORTH AMERICAN NONDESTRUCTIVE TESTING SOCIETIES AND RELATED ORGANIZATIONS

| | |
|---|---|
| Canada | Canadian Society for Nondestructive Testing (CSNDT) |
| CEN | European Committee for Standardization |
| China | Chinese Society for Nondestructive Testing |
| France | Confédération Française pour les Essais Non Destructifs |
| Germany | Deutsche Gesellschaft fur Zerstorungsfreie Prufunge (DGZfP) |
| ISO | International Organization for Standardization (ISO) |
| Russia | Russian Society for Nondestructive Testing and Technical Diagnostics (SSNTTD) |
| United Kingdom | British Institute of Non-Destructive Testing |
| United States | American Society for Nondestructive Testing (ASNT) |

## ► EVALUATING AND INSPECTING HEAT EXCHANGERS

### Overview

Internal inspections are a major undertaking and must balance two opposing factors

   Need to investigate poor performance (wasted energy) or mechanical failure concerns

   High cost of the shutdown and inspection required to fully address performance and failure concerns

Driven by API 510 inspection requirements

Basic decision process

   Open? (must clean if open)

   Inspect/Clean/Inspect

   "OK" or "Not OK"

Repair

Change out?

One method to optimize scheduling of exchanger cleaning is tracking

Tracking procedures are in place to optimize two separate programs

Process monitoring

Inspection requirements

Don't leave the inspection up to the inspector

They are interested primarily in mechanical integrity, not process optimization

Engineers rely on inspectors for feedback on how the exchanger is holding up

Additional information can be gained during an inspection to make the exchanger perform better and longer

### How to Evaluate a Heat Exchanger

Evaluations use inspection results to develop a plan of action

Focus of evaluation is to make a decision on

Repair or replacement

When best to do this work

"Inspecting" is the actual examination of the exchanger, while

"Evaluating" is taking inspection results together with all the other information concerning the exchanger to decide how best to optimize performance and reliability of the exchanger by

Cleaning

Repairing

Replacing

The standard approach to solving the problem is

Identify the problem

Gather information, including inspection results

Analyze the information
Formulate conclusions and recommendations
Have others check your work
Take action

## How to Inspect a Heat Exchanger During Shutdown

Study heat exchangers to be opened prior to shutdown
  Identify service
  Review service history
  Study drawings and design data sheet
  Visualize operation and potential problems
  Have a contingency plan in case cleaning doesn't work or
    walls are too thin
Know why heat exchanger is being opened
  Scheduled inspection
  Poor thermal performance
  Excessive pressure drop
  Suspected leak
General inspection
  Overall condition
  Configuration match drawings
  Look for and describe fouling or debris
  Look for and describe metal loss
Fouling and debris
  Appearance (description or photo for records)
  Location (drawing or photo for records)
  Collect representative sample
  Visualize mechanism
Metal loss
  Vibration or corrosion (description or photo for records)
  Location (drawing or photo for records)
  Visualize mechanism

Shell-and-Tube Heat Exchanger Cleaning Techniques

Washing

All cleaning methods are preceded by washing

Remove acids (by soda wash), oils, and toxins

Purpose is to make the exchanger safe for opening and entry, not to remove fouling deposits

Hydro-blasting can

Primary cleaning method for shell-and-tube exchangers

Effective for both shell-side and tube-side cleaning

Chemical cleaning

Goals

Remove heavy oils, light oils, benzene, pyrophorics, and odor

Increase shutdown safety

Minimize environmental impact

Sequence

Start with steam, followed by a noncaustic soap and then a potassium permanganate oxidizer

Technique options

• Do it yourself

• Get help (i.e., Halliburton) or

• "Turnkey" contract the whole job (i.e., Serv-Tech)

Effectiveness for improving heat transfer is questionable.

Mechanical reaming of tubes

Can only be performed for straight tubes

Seldom chosen as a cleaning method

After the bundle is removed from the shell

Must be moved to an isolated area before it can be cleaned

Cleaning area must provide for proper disposal of the wastewater from cleaning

Can the heat exchanger be cleaned on the run?

Possibly; service specific

## Inspection Techniques

Inspection/engineering each needs data from the other

Inspector

Focus is different than engineering

Ensure mechanical integrity

Looks for signs of

Leaks/cracks

Corrosion/erosion

Fatigue

Warpage or distortion

Mechanical failure or impending failure

Also check and documents

Tube thickness

Gasket surface condition

Tube-to-tube sheet connection integrity

Rely on nonvisual inspection techniques

Remote field (eddy current testing)

Radiography

Pressure testing

Need historical data and U values from the process engineer

Network heavily with other facilities that have similar exchangers

Inspection results data are not as meaningful as it would seem

## Bundle Replacement (Retubing)

Consider all other options before retubing an exchanger

There is a mechanical limit on the number of times a tube sheet can be retubed

Want to fully utilize the life of each exchanger so use retubing only as a last resort

When to retube

Potential penalties for tube failure are too high

Plugging tubes would exceed allowable limit of 10% of total number of tubes plugged

Rerolling has been ruled out as a solution

Operational change has been ruled out as a solution

No other cost-effective options exist

Retubing decisions are much easier if exchanger historical data are available to consult

Retubing takes a lot of planning; project tasks include

Determining if redesign is necessary

Choosing metallurgy

Considering a conversion to a U-tube bundle

Redesigning the tube bundle

Challenges

Spotty data

Making good business recommendations

### Plugging off a Leaking Tube

Plugging tubes is a business decision

Based on the sacrifice of a small measure of exchanger performance in return for prolonged exchanger life

Important to clean the inside of tubes before plugging

Need to replace badly worn or impinged tube ends with a new tube or a short tube plug before inserting a tapered plug

Location matters

Distribution of plugged tubes within the bundle will affect the total number of tubes that can be plugged before needing retubing/replacement

Look for any patterns that help determine the root cause of tube leaks

Develop a "plug map" for equipment records and future troubleshooting

Sometimes a leaky tube can be replaced instead of plugged if the location permits

## Impingement Protection Repairs

Investigate the root cause of impingement protection damage

Location of impingement protection devices makes cut-and-replace or shaping repairs a straightforward matter

Recommend removing all impingement protection devices that restrict flow in or out of the nozzle

These devices may have been placed in the piping or in the nozzle

When getting a new bundle, consider changing to impingement rods from impingement plates

## Torqueing Procedures (Bolt Up)

Not as straightforward as it seems (refer to Figures 3.1 and 3.2). Each heat exchanger has special requirements

Excel spreadsheets are available to help calculate torque values (Refer to Figures 3.6-3.8)

## Gasket Details

Graphite-coated corrugated metal gasket (Figures 3.3-3.5)

## Baffle Repair and Seal Strip Replacement

Options are limited due to access constraints

The leaf seal is made of thin sheets of high alloy metals

Leaf seals shall be replaced every time bundle is removed

We do not bend many baffles

If there is any significant damage to the baffle you must retube or replace the bundle

## Gasket Surface Repair and Requirements

Consider the gasket and flange surface as a system (Figure 3.9)

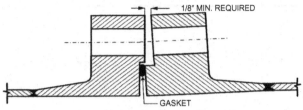

Rotation reduces effective seating area of the gasket. The joint is probably not leaking because the fulcrum for rotation is the gasket

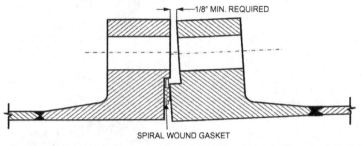

SPIRAL WOUND GASKET

The fulcrum of rotation is the compression in the spiral wound gasket. Increase in the bolt stress will decrease the gasket stress.

**Figure 3.1** Impact of flange rotation on gasket seating.

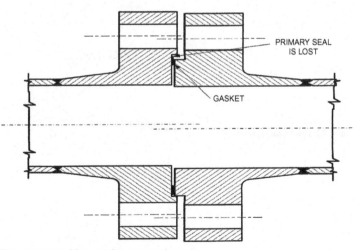

**Figure 3.2** Impact of flange misalignment on gasket seating.

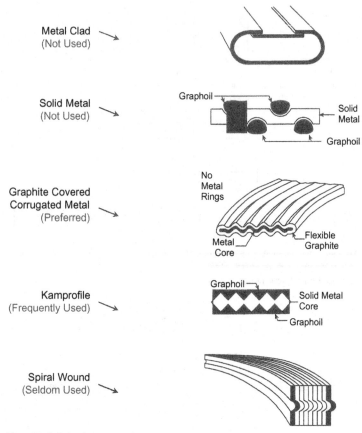

**Figure 3.3** Typical gasket types.

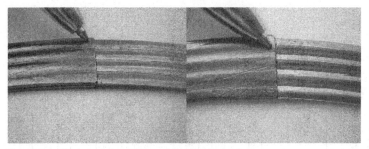

**Figure 3.4** Graphite is removed on half the gasket to show the metal core.

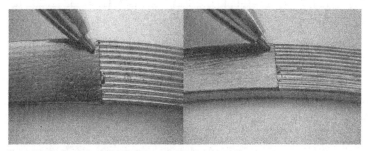

**Figure 3.5** Graphite is removed on half the gasket to show the metal core.

Conditions requiring repair
  Too rough or too smooth
  Warped
  Corroded
  Gouged or scratched
Before skim-cutting (resurfacing) a flange
  Flange meets ASME minimum thickness requirements
    Ensure not too thin
  Weld buildup or skim cut
  Final surface finishing to required roughness/smoothness

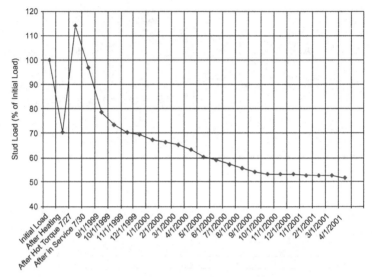

**Figure 3.6** Stud load changes due to short-term and long-term gasket relaxation.

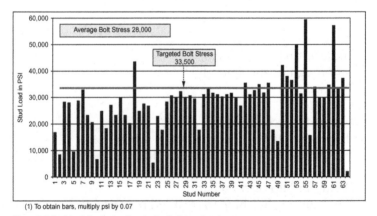

(1) To obtain bars, multiply psi by 0.07

**Figure 3.7** Stud load distribution for wire-brushed reused studs.

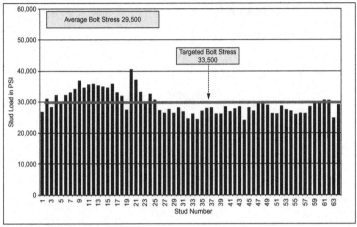

(1) To obtain bars, multiply psi by 0.07

**Figure 3.8** Stud load distribution for new studs.

Monitor the remaining thickness of the flange weld repairs;
   also have the danger of being too rough or too smooth
Plywood protectors are a good idea for protecting gasket
   surfaces during turnarounds

## Tube Impingement

Impingement problems are normally corrosion/erosion
   problems
   Primary concern is tube sheet joint impingement (high
      tube-side velocity)
   High energies are involved in erosion—local fluid veloc-
      ities of over 100 ft/s (30.48 m/s) are needed
   Consider channeling and metallurgy effects on the situation
Solutions
   Change metallurgy
   Eliminate corrosion by not mismatching materials
   Mitigate the chemistry causing the corrosion
   Put some rods in to be impinged (sacrificial anode
      philosophy)

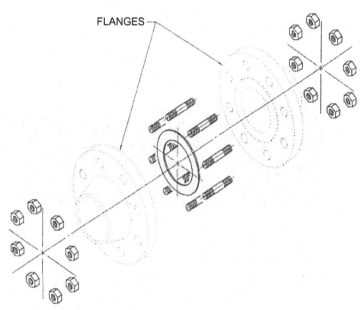

**Figure 3.9** Gasket and flange system.

## Retrofitting to Prevent Tube Vibration

Retrofitting is easier with square pitch tube sheets (Figure 3.10)

Vibration is usually stopped by mechanical supports

  Change the harmonic characteristics of the bundle to stop vibration

  Consider leaving the bundle as is and jam in additional rods to "detune" the bundle while adding additional tube support

Work can be performed without taking the tube bundle apart

In the design stage, try to block a leakage stream

  Flow around outside of bundle (C-stream leakage)

  Flow into a partition lane

Figure 3.11 shows the rotated square pitch diagram

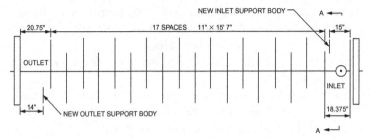

**Figure 3.10** Retrofit tube supports—Plan view.

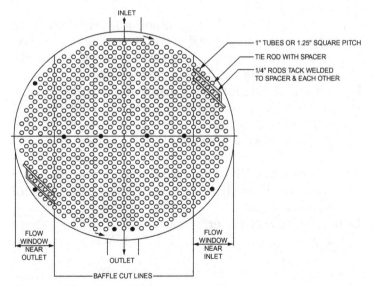

**Figure 3.11** Retrofit tube supports—Elevation view.

Tube vibration prevention action is taken for carefully selected tubes based on localized flow velocity generated
by a computer program for velocity
There is a definite distinction between triangular and
square pitch
   For triangular pitch bundles, because tube lanes are small,
     rods inserted into the tube bundle can cause the tubes to
     bend if not sized and installed properly
   For square pitch bundles, tube lanes allow easy insertion
     of bar-stock to stop vibration

**Hydro Testing**

The pressure test is an important final step of maintenance/
repair actions
For the pressure test following **minor** heat exchanger work
   The lower test pressure value corresponding to the max-
     imum allowable working pressure (MAWP) lessens the
     likelihood of gasket damage and leakage
For pressure testing following **major** modifications involv-
ing through wall welding on a pressure boundary
A hydrotest to
   1.5 times MAWP is required for equipment built to a
     pre-1999 ASME Code corrected for temperature
   1.3 times MAWP is required for equipment built to a
     post-1999 ASME Code Addendum corrected for
     temperature
Example temperature corrections:
Test pressure in psi (MPa) = 1.5 MAWP $\times$ ($S_{test\ temp}/S_{design\ temp}$)
Test pressure in psi (MPa) = 1.3 MAWP $\times$ ($S_{test\ temp}/S_{design\ temp}$) 1999 addendum and later
where

$S_{\text{test temp}}$ = allowable stress at test temperature in ksi (MPa)

$S_{\text{design temp}}$ = allowable stress at design temperature in ksi (MPa)

## ▶ TUBULAR EXCHANGER INSPECTIONS

### Overview

Exchangers are a very important part of any operation and are found in almost all process units

There are many types of exchangers, but by far the most common are tubular exchangers, also referred to as shell and tube

The remainder of this section covers

Field and shop inspection and repairs

Minor welded repairs to complete rebuilds

Most steps of exchanger shop work will be covered

Initial cleaning

Visual inspection

Nondestructive examination

Destructive examination

Minor repairs

Major repairs

Pressure testing

Alterations

Quality control inspection tools

## ▶ INITIAL EXCHANGER INSPECTION

### Procedure

The first step in bundle inspection is a general visual inspection that may establish general corrosion patterns

Bundles should be checked when they are first pulled from the shells

Color, type, amount, and location of scales and deposits often help to pinpoint corrosion problems

An overall, heavy scale buildup on steel tubes may indicate general tube corrosion

Lack of any scale or deposit on tubes near the shell inlet may indicate an erosion problem

A green scale or deposit on copper base tubes indicates that these tubes are corroding

As an inspector gains experience, these scales and deposits will become a useful inspection guide

While inspecting a bundle visually

The inspector should make use of a pointed scraper to pick at suspected areas next to tube sheets and baffles

These areas may not have been cleaned completely

Picking in these areas will sometimes disclose grooving of tubes and enlargement of baffle holes

Tapping the tubes with a light [4- to 8-ounce (115- to 225-g)] ball peen hammer or inspection hammer during the visual check

Helps locate thinned tubes

Useful when inspecting light-wall tubes of small outside diameters

The amount of rebound and the sound of the blow give an indication of the tube wall thickness

This method becomes more helpful as experience is gained in the use of the hammer

Inside of the tubes can be partially checked by the use of

Flashlight extensions

Fiber-optic scopes

Bore scopes

Special probes are

Slender 1/8-in. (3.2-cm) rods with pointed tips bent at 90° to the axis of the probe

Possible to locate pitting and corrosion near the tube ends

Only the outer tubes of a bundle can be thoroughly inspected externally and without a bore scope or fiber-optic scope; only the ends of tubes can be inspected internally

Complete inspection of the tubes for defects or thickness is accomplished using

Eddy current or ultrasonic methods

**Procedure—Destructive Inspection**

Tubes may also be removed from the bundle and split for visual inspection

Devices are available for pulling a single tube from anywhere in a bundle (refer to Figures 3.12 and 3.13)

A cutting tool is inserted into the tube as depicted in Figure 3.12 and then rotated by a pneumatic tool

Removal of one or more tubes at random

Permit sectioning

More thorough inspection for determining the probable service life of the remainder of the bundle

In the case of this design of exchanger, the type tube extractor may be the only way to remove tubes from the exchanger bundle while saving the tube sheet for rebuilding the exchanger

This is a painstaking task; this method is only used with a design that will not allow removal of tubes by the preferred method of cutting the tube sheet from the bundle and driving the tubes out of the tube sheet from the reverse side

When select tubes are removed for inspection purposes and it is determined that no repairs need to be done, the empty tube sheet holes are plugged

**Figure 3.12** Collet-style tube tugger pulling bars.

**Figure 3.13** Collet-style tube tugger pneumatic tool.

This is done using tapered plugs made from the appropriate material for the service

In most cases, installing replacement tubes is not required and plugging is the usual method to make the exchanger serviceable

Tube removal is also employed when special examinations, such as metallurgical and/or other evaluations, are needed; some examples are

Dezincification of brass tubes

Depth of etching or fine cracks

High-temperature metallurgical changes

In Figure 3.14, a tube from an air-cooled exchanger has been split for examination

**Figure 3.14** Air cooler tube with severe pitting.

## Procedure—Nondestructive Inspection

When bundles are retubed, similar close inspection of tubes removed will help identify the causes of failure and improve future service

Baffles, tie rods, tube sheets, and a floating-head cover should be inspected visually for corrosion and distortion

Gasket surfaces should be checked for gouge marks and corrosion and other forms of damage

A scraper will be useful when making this inspection

A sufficient gasketed surface should remain to make a tight seal possible when the joints are completed

Tube sheets and covers can be checked for distortion by placing a straightedge against the flat surface and measuring the outside dimensions

Distortion of tube sheets can result from

Overrolling or improper rolling of tubes

Thermal expansion or explosions

Rough handling or overpressurization during a hydrotest

Tube sheet and floating-head thickness can be measured with mechanical calipers or Ultrasonic A Scan thickness meters

Except in critical locations, continuous records of such readings are not usually kept. However, the original thickness readings of these parts should be recorded

Thickness readings of tie rods and baffles are not generally taken

Condition of these parts is determined by visual inspection

Tube wall thickness should be measured and recorded at each inspection

It is sufficient to measure the inside and outside diameters and to thus determine wall thickness by calculation

Figure 3.15 is an example of measuring tube ODs; this type
of outside caliper is available in several sizes from small to
quite large

It is obvious that using the inside and outside measurement
method for calculating tube wall thickness is restricted to
tubes near the tube sheet, as can be seen in Figure 3.16,
and outer tubes in the bundle

Eccentric corrosion or wear noted during visual inspection
should be taken into account in determining the remaining life of the tubes

Several tools are available for the assessment of tube
conditions

Long mechanical calipers can be used to detect general or
localized corrosion within 12 in. (30.5 cm) of tube ends

More detailed measurements along the entire tube length
can be achieved with specialized tools, such as

**Figure 3.15** OD tube measurement.

**Figure 3.16** Inside diameter measurements.

Laser optical devices
Internal rotary inspection system (IRIS)
Electromagnetic techniques such as eddy current (EC)
Laser optical and ultrasonic devices require a high degree of
  internal tube cleanliness compared to electromagnetic
  methods
Laser optical devices can only detect and measure internal
  deterioration
Electromagnetic methods (EC)
  Can detect and provide semiquantitative information on
    both internal and external defects
  Cannot determine whether the defects are on the internal
    or external surface of the tubes
Rotary ultrasonics (IRIS) will generally provide
  Most quantitative information

Can identify if defects are on the internal or external surface of the tube

The aforementioned methods are better described later in
this chapter, but as we will see all methods have limitations and which is most appropriate is important information in obtaining credible data on which to make
repair/replace decisions

The choice of technology is largely based on the industry
experience of experts in a particular system of inspection

When an assembled exchanger is shipped to a repair facility,
which is not usually the case, inspection personnel may
perform an external inspection prior to tear down for extensive internal inspection

One primary such nondestructive inspection is to take part
thicknesses using a digital thickness meter on areas such
as the exchanger shell and closures

In Figure 3.17, the quality control department is measuring
an exchanger shell wall thickness using an ultrasonic digital meter

Thickness of nozzles and other parts will also be taken and
entered into a exchanger evaluation report for the owner–
user's review

When working with alloy materials, there should be confor-
mation of material chemistry before repairs commence
involving welding or replacement; the most effective
way to accomplish this is to use a method of positive material identification (PMI)

By far the most common field method used for PMI is
readings taken directly from the alloy part using a
hand-held device, the hand-held nuclear analyzer

Figure 3.18 shows readings being taken using this device on
stainless steel exchanger tubes; the analyzer identified the
tubes to be alloy 304 stainless steel

**Figure 3.17** Shell thickness readings.

**Figure 3.18** PMI of stainless steel alloy.

## ▶ MOST LIKELY LOCATIONS OF CORROSION

### General Considerations

Locations where corrosion should be expected depend on the service of the equipment

There are certain locations that should be watched under most conditions of service

The outside surface of tubes opposite shell inlet nozzles may be subject to erosion or impingement corrosion

When a mildly corrosive substance flows on the shell side of the tube bundles, maximum corrosion often occurs at these inlet areas

The next most likely point of attack under the same conditions would be adjacent to baffles and tube sheets; any deterioration here is probably erosion–corrosion (Figure 3.19)

**Figure 3.19** Erosion/corrosion shell inlet.

When a high temperature material flows into the tube inlet pass, the backside of the stationary tube sheets or tubes immediately adjacent to it may suffer extensive corrosion

When process conditions allow a sludge or similar deposit to form, it will generally settle along the bottom of the shell

If the deposit contains a corrosive material, maximum corrosion will occur along the bottom of the shell and the bottom tubes

In water service, maximum corrosion will occur where the water temperature is highest

Thus, when water is in the tubes, the outlet side of the channel will be the location of maximum corrosion (Figure 3.20)

In water service, when exchanger parts are made of gray cast iron, they should be checked for graphitic corrosion

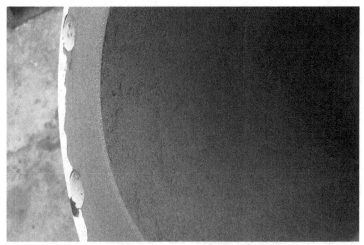

**Figure 3.20** Pitting in a channel shell.

## Graphitic Corrosion

Is the deterioration of gray cast iron in which the metallic constituents are selectively leached or converted to corrosion products, leaving graphite in place of the metal?

Sometimes also referred to as **graphitization**

Most often found in water-service channels or along the bottom of shells where sour water might collect

Found by scraping at suspected areas with a stiff scraper

## Sour Water

Contains sulfur compounds (usually hydrogen sulfide) at concentrations of 10 ppm by weight or more

Seriousness of attack depends on its location and depth

  Quite often, pass partitions can be almost completely corroded and still function efficiently, unless the carbon shell is broken or chipped

In any type of exchanger, corrosion may occur where

  Dissimilar metals are in close contact

  Less noble of the two metals will corrode

## General Considerations

Carbon steel channel gasket surfaces near brass tube sheets will often corrode at a higher rate than they would otherwise

Cracks are most likely to occur

  Where there are sharp changes in shape or size or

  Near welded seams, especially if high stress is applied to the piece

Parts such as nozzles and shell flanges should be checked for cracks if excessive stresses have been applied to a unit

When process stream velocities are high in exchangers

  Erosion damage can be expected at changes in the direction of flow

Damage would occur on or near such parts as tube inlets
in tubular units and at return bends in double-pipe units
and condenser box coils

The area of the shell adjacent to inlet impingement plates
and bundle baffles is susceptible to erosion, especially
when velocities are high

A distinctive Prussian blue color on bundle tubes indicates
the presence of ferri-ferrocyanide

Hydrogen blistering is likely to be found on the exchanger
shell near a Prussian blue color

A long straightedge may prove useful in determining the
existence of blistering

### Reference: Material on Damage Mechanisms

The American Petroleum Institute's Recommended Practice
571, entitled "Damage Mechanisms Affecting Fixed
Equipment in the Refining Industry," is one recent reference document regarding corrosion and other types of
damage to fixed equipment

API RP 571 is an excellent reference document for determination of what type of damage might occur based on the
service, as it details the locations in equipment where a
particular type of damage might be present

It also provides color photographs of the different damage
mechanisms

Lists where in a refinery, meaning which units to expect a
particular type of damage

More specifically, where in an exchanger or vessel to inspect
for damage

RP 571 also details inspection methods and nondestructive
examinations, which are most effective for revealing and
assessing the anticipated damage

## ▶ SHOP WORK

### Initial Cleaning

Cleaning exchanger bundles and shells may be performed in the plant or repair facility

Often only the tube bundle and possibly some exchanger parts, such as channel and heads, are sent to the exchanger shop to be cleaned, inspected, and repaired as needed

Sometimes the entire exchanger assembly is transported to the repair concern, but costs and environmental concerns often limit this approach unless the repair facilities are properly equipped to dispose of harmful chemicals

Some shop repair facilities have advanced cleaning equipment with wash bays for cleaning both the outside and the inside of tube bundles and can handle most contaminants safely

### Cleaning Bays

Usually the ODs of the tubes are hydroblasted followed by ID blasting (referred to as lancing)

This cleaning process

   Can take up to 12 hours or more

   Depends on the hardness of deposits on the outside of the tubes

Tube bundles from a crude oil unit, for example (Figure 3.18)

   May require a much longer time in the bay

   If the bundle is a retube after decontamination, it goes to the shop area from OD blasting

### Crude Unit 5 Chrome Bundle Example

The exchanger bundle shown in Figure 3.21 was in for retubing so initial cleaning was not required

Figure 3.21 Crude unit bundle.

Had cleaning been required, it would have taken a very long
   time and some aggressive chemicals
It might seem that this bundle should have been scrapped,
   but all the internal parts, such as baffles and tube supports, are made of a 5Cr alloy, which makes the parts very
   expensive and worth salvaging
Only the tubes were replaced and some machining was performed on this exchanger
A great deal of deposit removal was needed to free the tubes
   for extraction
Once the bundle cage is bare, then normally grit blasting of
   the parts is used for cleaning and prior to inspection

## OD Blasting

The entire tube bundle is placed inside an OD cleaning bay on a moveable rail car; the rail car is designed to roll the bundle as it is being cleaned

The operator will monitor the cleaning progress through an observation window; once it is determined that a segment of the bundle has acceptable cleanliness, it is rotated to begin cleaning of the next segment

As mentioned previously, this can take a few hours or more

Sometimes cleaning is halted after one 12-hour shift, as additional cleaning is not effective; however, it can be extended well beyond a single 12-hour shift

One such OD cleaning bay is shown in Figure 3.22

**Figure 3.22** 10,000-psi (690-bar) OD blasting.

Acceptance by the owner–user at the achieved cleanliness or grit blasting to try to improve cleanliness to specification

## ID Blasting (Lancing)

When a bundle is inspected, it goes from OD to ID blasting (lancing)

Depending on the hardness of deposits, lancing can take 12 hours or more to get the inside of the tubes clean

Some tubes may be plugged so severely with product deposits that blasting will not clean the tubes

In such cases where product blockage is severe, then plugging and drill venting the affected tubes can be done if

They are few in number; about 10% of tubes can be plugged without affecting an exchanger's performance

Figure 3.23 shows the machinery layout used for this purpose

All of the wash facilities drain to a wastewater recovery system

## Grit Blasting

When needed, tube bundles are transported from the hydroblasting bay to the grit bay for grit blasting

Internal tube cleaning is normally done only when this level of cleanliness is required for nondestructive testing methods such as the IRIS

Figure 3.24 shows a grit-blasting chamber

Grit blasting must be done by a competent individual, as damage can occur if this procedure is not performed properly

Damage to the bundle can result in additional costs up to and including a complete rebuild or replacement of the tube bundle

**Figure 3.23** 10,000-psi (690-bar) ID blasting (lancing).

**Figure 3.24** Grit chamber used to blast tube bundles.

The type of grit used for blasting will depend on the level of cleanliness required; normally commercially available sand is used

As shown in Figure 3.25, the two tubes circled in white were cut by very hard and aggressive grit referred to as "Black Beauty"

The exchanger bundle of Figure 3.20 is of the double tube sheet design

As such the need for repair due to the grit blasting operator's mistake was a major concern

Double tube sheets are called for when the design of the exchanger requires that the two fluids used in the exchanger

Must not come into contact as the result of an in-service leak at tube rolls

**Figure 3.25** Tube end damage from grit blasting.

The concern might be
  Violent chemical reaction
  Loss of an expensive product due to contamination
  Accelerated corrosion upon mixing of the two fluids
This type of damage to any equipment is always a concern
For this more complex design it will be many times more
  expensive to repair than normally required
This damage was judged superficial and as such it did not
  require the difficult and expensive repairs that would
  be needed had damage been extensive
This damage was completely avoidable; careless work re-
  sults in unnecessary repairs and can jeopardize the integ-
  rity of an exchanger
Even what seems to be a low-tech endeavor must be done
  correctly or mistakes like this will occur
Luckily, things like this are rare, as they should be

## ▶ SHOP INSPECTION

### Clean Visual Inspection

Once the bundle has been cleaned, a visual inspection can
  be performed to determine if the
  Bundle can be placed back in service
  May need some form of repair, up to and including com-
    plete rebuilding
Some items to inspect include
  General tube bundle condition, such as mechanical dam-
    age in the form of severely bent baffles or tubes
  Tube ends for thinning, corrosion, cracks, or severe me-
    chanical damage
  Tube sheets for warping, and damage to gasket seating
    areas
  Baffles and baffle tube holes for damage or corrosion

Sealing strips if so equipped for damage or corrosion
Tie rods loose, nuts missing, corroded or broken rods
Inlet impingement plate(s) condition

### Inspection of Exchanger Bundles
In Figure 3.26, some of the tube ends have impact damage;
however, these were leak tight during pressure testing, as
a result the damaged tubes were determined to be fit for
service
Due to the method by which tubes are expanded into the
tube sheet, a tight seal is made
Tube holes in the tube sheet contain what are referred to as
serrations
Serrations are narrow grooves machined into the internal
circumference of the tube sheet holes; once expanded
properly, the tube joint is sealed by expansion of the
tube wall metal into the serrations

**Figure 3.26** Tube end mechanical damage.

The term rolling tubes is the act of positioning the tubes in the tube sheet and using a rotating tool to expand the tube wall into the serrations

### Tube Sheet Groves

When a tube is expanded (rolled)
   The tube wall expands into the grooves (Figure 3.27)
   Providing a tight seal
   Grooves can be seen and felt as indentions a short distance inside the tube
Damage to the tube ends shown in Figure 3.21 was
   Outside of the tube sheet
   Did not extend into the tube sheet holes

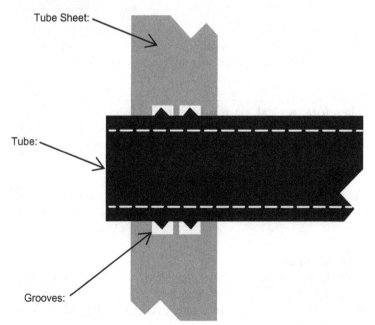

Tube Sheet:

Tube:

Grooves:

**Figure 3.27** Tube sheet grooves.

**Figure 3.28** Admiralty Brass tube sheet with tube groves in serviceable condition.

Therefore, no leak path is present

Figure 3.28 shows an Admiralty Brass tube sheet with tube
  roll grooves in serviceable condition, meaning
    Tubes holes are within specification for roundness
    The internal tube hole grooves are usable
After cleaning and measurements, this tube sheet was used
  to rebuild the exchanger


**Tube Hole Damage**

Tube holes are sometimes found with mechanical damage
  or egg shaped (out of roundness)
    Note the damaged tube holes marked with a white X in
      Figure 3.29

**Figure 3.29** Admiralty Brass tube sheet with tube-hole damage.

In Figure 3.29 there has been mechanical damage to the marked tube holes, which most likely occurred during the removal of old tubes in preparation for retubing the exchanger

With only a few tube holes damaged, the holes can be welded and machined back to dimension; the tube hole serrations will also be restored

If tube holes in the sheet are free of mechanical damage, the next step is to measure the diameter of the tube holes and determine

    If too large, making a tight roll within specifications is likely unobtainable

    If internal serrations (grooves) are in good condition

Figure 3.30 shows a specialty digital ID micrometer suited for this purpose

**Figure 3.30** Inside diameter micrometer.

## Tube and Tube Sheet Corrosion

Figure 3.31 reveals tube and tube sheet corrosion

    The tube internal surfaces were also severely corroded and pitted

    The severely corroded tubes circled in Figure 3.31 have corroded inward until the tube ends are inside the tube sheet and were probably leaking when removed

    This bundle was rebuilt almost entirely; all that could be salvaged were the tube cage and tube sheets

When dissimilar metals are used in combination with a fluid environment that will promote dissimilar metal corrosion (galvanic corrosion) and no special precautions are taken, this type of corrosion is likely to occur

In Figure 3.32, a carbon steel tube sheet, equipped with stainless steel tubes, has suffered very aggressive dissimilar metal corrosion

**Figure 3.31** Tube and tube sheet corrosion.

**Figure 3.32** Dissimilar metal corrosion.

Note the corroding of the tube sheet around the uppermost tubes; eventually this tube sheet would have been completely destroyed and would likely leak once it approached the second tube sheet serrations as by then the tube seals will be compromised.

The shell of the exchanger in Figure 3.27 is also made of carbon steel and has the same dissimilar metal corrosion damage as the tube sheet

It would be a good plan to design and build the replacement using 100% stainless steel components for all wetted surfaces exposed to the exchanger's chemical environment, assuming that stainless steel is, in fact, the correct metallurgy for this application

Figure 3.33 shows a typical layout before cutting out shell segments to expose tubes for inspection and a possible retubing in this fixed tube sheet exchanger

**Figure 3.33** Layout for cutting off shell.

There were several of this type of fixed tube sheet exchangers in the shop for repair at the same time; most were repairable, some were not

The exchanger in Figure 3.34 has been retubed and the two halves of the shell, previously cut off to provide access to the tubes, have been replaced, fitted, and tack welded in preparation for final welding out, which will complete the repair

### Shell Corrosion

Figure 3.35 shows damage to the carbon steel exchanger shell. This shell was cut away initially with the intent of retubing the exchanger; once exposed and the

**Figure 3.34** Shell halves tack welded in place.

**Figure 3.35** Corroded shell.

extensive damage became known, it was obvious that repair was not feasible

Tube sheets shown previously in Figure 3.35 are scrap;
tubes will be cut from it, making the tubes too short for
reuse; the shell also has extensive internal corrosion

While the initial cost of this particular exchanger's design
was relatively inexpensive, its cost of replacement, loss
of capital equipment, and possible extended downtime
of its production unit may have made it a very expensive
design choice

### Added Metal Thickness for Corrosion Allowance

How much extra metal used for corrosion allowance in a
particular design is relative

Most codes normally do not have a mandatory amount
stated, instead a code might make the comment that

corrosion allowances should be accounted for in the design of equipment when used in a corrosive service

The designer will decide the amount of corrosion allowance based on the fluid service, material chosen for construction, and owner–user requirements for life expectancy

Plain carbon steel, in many instances, will require only a 0.125-in. (3 mm) allowance

One criterion in design is the required life expectancy of the exchanger shell or other type of pressure equipment

A standard industry practice is to design for 25 years as the least usable life span for many vessels

Using tables from the International Association of Corrosion Engineers (NACE), designers can look up the expected average corrosion rate for most fluid and material combinations

The NACE tables list the various fluids, usually in a temperature percent concentration format and give the approximate corrosion rate per year that is to be expected

Should an estimated corrosion allowance be excessive, then the choice of material would need to be reconsidered, for the required service life desired

### Damage to Tube Sheet Gasket Area

Figure 3.36 shows severe gasket seating surface corrosion

Damage is due to a leaking gasket between the tube sheet and the body flange of the exchanger shell

The body flange of the shell may also be corroded

Unless other severe damage exists, such as enlarged tube sheet holes, this tube sheet and shell body flange may be machined back to tolerances; this may require weld metal buildup prior to machining

Prior to weld metal buildup, a light machine cutting pass is often used to ensure sound metal for weld and base metal fusion

**Figure 3.36** Corroded gasket seating area.

In some instances, the only proper repair is to manufacture a replacement tube sheet

Figure 3.37 reveals mechanical damage to the tube sheet gasket

This is typical of the type of damage that occurs due to improper handing

This damage was repaired by weld metal buildup and machining back to specifications—had more care been taken, this repair would not have been required

This type of damage usually occurs when channel covers are being removed or reinstalled in the field, but again this is a result of handling

**Figure 3.37** Mechanical damage to gasket seat.

## Enlarged Baffles Tube Support Holes

Baffles in an exchanger serve three purposes

    Support tubes during assembly

    Direct fluid flow on the shell side to enhance heat transfer

    Support tubes during operation to keep vibrations to a
        minimum

Baffle tube holes in the transverse baffle, which are en-
    larged, as shown in Figure 3.38, may permit the tubes to
    vibrate excessively, causing fatigue cracks, or the wearing
        holes through tubes causing leaks; the only proper
        repair is replacement of the baffle

On occasion, an operating exchanger may emit an audible
    humming sound

**Figure 3.38** Enlarged baffles/tube support holes.

This sound may be due to
  Tubes vibrating in enlarged tube holes
  A sign leaks may be present or imminent
The only cure is retubing the bundle using new
  Baffles and tube supports
  Tie rods
  Tube sheet if it is damaged
Sometimes new tube bundles are prefabricated in advance
  of a shutdown based on
  Inspection records indicating the bundle was found to be
    marginal during the last inspection cycle
  Internal leaks in tubes that are indicated by process fluid
    contamination
Exotic metal alloy bundles may have long lead times, neces-
  sitating the need for prefabrication

Such bundles can be prefabricated and stored at the owner–user's facility well in advance of installation

### Seal Bars and Longitudinal Baffles

The two short horizontal bars shown in Figure 3.39 are leaf seal bars used to

Direct fluid path on the shell side

They are a form of a longitudinal baffle and are used in conjunction with transverse baffles to direct fluid flow on the shell side

On this exchanger, there are four such leaf seal bars spaced 180° apart at opposite ends of the bundle

Bars may be placed elsewhere; there are several different designs

**Figure 3.39** Leaf seal bars.

Used for operational efficiency, helping direct fluid flow on the shell side of the exchanger to minimize fluid bypassing the intended path

Often they are bent during extraction and insertion of the tube bundle

They are relatively inexpensive and are normally replaced if they cannot be brought back into position

Another design for shell-side sealing is one made of a plate running the length of the exchanger bundle

They can run the entire length of the tube bundle or only a portion of its length

Depending on the fluid flow path chosen for the shell side of the exchanger, long or short baffle plates are installed

Figure 3.40 is an example of one such configuration using a plate

As with all parts of the exchanger bundle, these must be inspected for thinning and for cracked welds

**Figure 3.40** Plate longitudinal baffles.

## Tie Rods

Figure 3.41 shows bundle cage (or skeleton) threaded round bars; these are tie rods that align and hold the vertical baffles. There are protective shrouds over each bar; these shrouds are not for strength but instead protect against erosion

They support the tube baffles during fabrication

They are used to fix the length of the assembly from end to end

Tie rods pass through the shroud shields between the baffles to space the baffles properly

During inspection of a bundle that has been in service, tie rods are checked for damage, such as bending, missing, or loose tie rod nuts and severe erosion or corrosion

Figure 3.41 Tie rods.

Normally, both ends of the tie rod have double nuts to assure they will not vibrate and come loose

### Inlet Impingement Plate

In Figure 3.42, the round plate is the impingement plate
  It is welded directly to a rigid support member such as the longitudinal bars shown in Figure 3.42
During inspection, the impingement plate may have broken welds, be very thin, or be missing altogether
Impingement plates opposite the shell inlet nozzle protect tubes from being eroded by the incoming fluid stream
When a shell has more than one inlet nozzle, there should be an impingement plate for each nozzle's inlet
Impingement plates are important and should not be overlooked

Figure 3.42  Inlet impingement plate.

## ▶ NONDESTRUCTION EXAMINATION

### General Considerations

Nondestructive examination of tubes is performed using
one of the following technologies
Internal Rotating Inspection System
Remote field eddy current (RFET)
Eddy current testing (ECT)
As mentioned earlier, IRIS requires a higher degree of tube
internal cleanliness than RFET or ECT

### IRIS Background

The IRIS system uses the ultrasonic pulse echo technique to
generate high-frequency sound waves into the tube wall
measuring the thickness over all scanned surfaces
Probes transmit an ultrasonic pulse, which is reflected at
right angles by the rotating mirror (45°) housed in the
probe turbine assembly (Figure 3.43)

### Applications

Field proven and used commonly in boilers, heat exchangers, and fin-fan tubes
Often used as a backup to electromagnetic examination of
tubes to verify calibration and accuracy. Especially useful

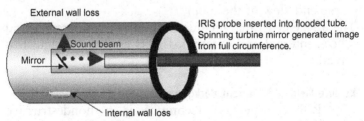

External wall loss

IRIS probe inserted into flooded tube.
Spinning turbine mirror generated image
from full circumference.

Sound beam

Mirror

Internal wall loss

**Figure 3.43** Probe turbine assembly.

as a follow-up to remote field testing due to full sensitivity near tube support structures provided by IRIS, which RFET lacks

## Limitations

The IRIS probe must be moved very slowly (approximately 1 in. per second, or 2.5 cm/s), but it produces very accurate results (wall thickness measurements typically accurate to within 0.005 in., or 0.13 mm)

Before examination, tubes must be cleaned on the inside to bare metal, referred to as white metal

A supply of clean water is needed, typically at a pressure of 60 psi, or 0.4 MPa; dirt or debris in water may cause the turbine to jam

Works for tube diameters of ½ in. (13 mm) and up; requires an adaptor for larger diameters

Works in metal or plastic tubes, typical smallest detectable defect: through hole of 1/16-in. (1.6-mm) diameter

Operates in temperatures above freezing; it can pass bends, but **will not** detect defects in bends and is **not sensitive** to cracks aligned with tube radius

Limitations of any method of nondestructive testing must be considered when choosing the method to be applied

The IRIS system produces a B scan or end view or cross-sectional image of the tube and can be viewed in a typical IRIS rectilinear format or circular view (representing a true end view of the tube)

The image generated includes full tube circumference, and both internal and external defects can be distinguished readily

## Remote Field Eddy Current Background

The RFET inspection technique is a nondestructive method that uses low-frequency AC and through-wall transmission to inspect pipes and tubes from the inside

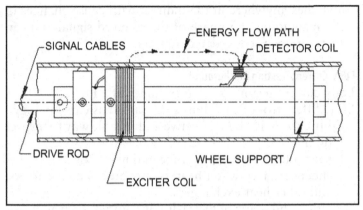

**Figure 3.44** Remote field eddy current.

The through-wall nature of the technique allows external and internal defects to be detected with approximately equal sensitivity

The RFET tool uses a relatively large internal solenoidal exciter coil, which is driven with low-frequency AC (Figure 3.44)

A detector, or circumferential array of detector coils, is placed near the inside of the pipe/tube wall, but displaced axially from the exciter by about two pipe/tube diameters

Two distinct coupling paths exist between exciter and detector coils

  The direct path, inside the tube, is attenuated rapidly by circumferential eddy currents induced in the tube's wall

  The indirect coupling path originates in the exciter fields, which diffuse radially outward through the wall

At the outer wall, the field spreads rapidly along the tube with little further attenuation

These fields rediffuse back through the pipe wall and are the dominant field inside the tube at remote field spacing

Anomalies anywhere in the indirect path cause changes in the magnitude and phase of the received signal and can therefore be used to detect defects

### Eddy Current Testing Background
Eddy current testing is the forerunner of RFET
ECT differs from RFET but is the same basic technology
  Eddy current is a cost-effective and reliable way to inspect tubing
  Inspect nonferromagnetic tube and materials
  Condensers, feed water heaters, air conditioners, chillers, and other heat exchangers
Detection and sizing of metal discontinuities, such as corrosion, erosion, tube-to-tube wear, pitting, fretting, and cracks
Multifrequency inspection with mixing and filtering capabilities

### Recommendations
The following approaches are recommended
  Nonferromagnetic materials, such as stainless steel, titanium, brass, Cu–Ni alloys, and Inconel
    Should be inspected by **eddy current testing**
    ECT has high detectability and high inspection speed for nonferromagnetic materials
  Ferromagnetic materials, such as **carbon steel**
    Can be inspected by **IRIS or RFET**
    IRIS should be used when **small pits** can be expected
When damage does not include small pits and is mainly general wall loss
  IRIS **or** RFET can be used
  IRIS will be **more accurate** but **slow** and requires significant cleaning
  RFET will be **fast and require minimal cleaning**

RFET works very well for inspection where the damage is
general wall loss

Equipment required is shown in Figures 3.45 and 3.46

In case of carbon steel tubes with **aluminum fins,** IRIS is
the preferred technique

The NDE examiner plays a significant role in the performance of these NDT techniques for tubing inspection

It is important that a performance demonstration be established to determine the ability of the examiner for

Detection of defects

Discrimination (valid defects vs false calls)

Sizing of all types of defects

ID, OD, small volume, large volume, cracks, and so on
should be inspected by eddy current testing

ECT has high detectability and high inspection speed for
nonferromagnetic materials

**Figure 3.45** RFET testing.

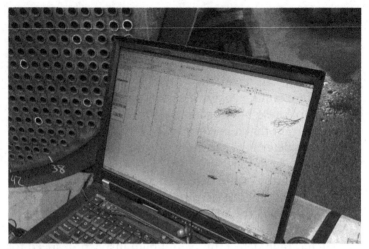

Figure 3.46 RFET display.

## ▶ MINOR REPAIRS

Examples of minor repairs

   Plugging tubes

   Minor weld metal buildup of gasket surfaces and machining

   Minor welding of tube sheets adjacent to tube ends

   Straightening of baffles

   Replacing worn or damaged seal strips

   Replacing tie rods and/or tie rod nuts

   Plugging tubes is quickly accomplished by

      Inserting a tapered plug often made of brass, but can be other materials

      Apply a few light blows with a hammer

   Minor weld repairs to tube sheets or seating surfaces, which consist of welding and resurfacing of the tube

sheet or gasket seating area by hand, are typical as
shown in Figures 3.47 and 3.48

Such minor repairs range from a few hundred to a few thou-
sand U.S. dollars depending on the man hours expended
by the repair facility

Often minor damage is not repaired; while the damage may
look bad, it does not affect the functionality of the exchanger and amounts to cosmetic work

In these cases the inspector might consult with the repair
concern and/or the owner–user and pronounce it fit for
continued service in its present condition

## ▶ MAJOR REPAIRS

Items considered major repairs include
    Complete retubing
    Retubing and manufacture of new baffles
    Retubing and manufacture of a new tube sheet
    The only other major repair to an exchanger would
        amount to building a new bundle where all of the fol-
        lowing are new
    Tubes and tube sheet
    Baffles and tie rods
    Seal bars, if so equipped

### Complete Retube (Tubes Only)

Retubing of a bundle begins with disassembly of the bundle
by cutting off the tube sheet(s) and removing the various
parts

This amounts to a six-step process as follows:
    Cut off tube sheet(s)
    Remove tube ends from the tube sheet(s)
    Remove tubes from the cage
    Disassemble the tube cage

**Figure 3.47** Weld buildup of tube sheet next to outer tube roll.

**Figure 3.48** Weld buildup of gasket seating area resurfaced with hand file.

Clean all parts (grit blast)

Inspect parts

Assembly of the refurbished bundle is a four-step process:

    Assemble the tube cage and tube sheet(s)

    Load the new tubes into the assembly

    Roll the tubes into the tube sheet(s)

    Leak test the refurbished bundle

The information that follows is an overview of the basic rebuilding process; as such it leaves out some steps and details

The first step in disassembly is cutting off the tube sheet as shown in Figure 3.49

**Figure 3.49** Cutting off tube sheets with a large-capacity band saw.

## Cutting off Tube Sheets

This saw's capacity allows removing a tube sheet from very
  large tube bundles (Figure 3.50)
  It can take many hours to complete the cutting operation
  This operation is the first step in rebuilding
After this operation
  The bundle is moved to a disassembly area for tube re-
  moval from the bundle cage
  The cage is inspected and damage noted for repairs
The tube sheet is moved to a separate area, and tube ends
  are driven out using a pneumatic hammer
Figure 3.51 shows a brass tube bundle that has an integral
  channel and tube sheet
  This type of tube bundle can have a different tube removal
  procedure than other designs

Figure 3.50 Band saw cutting a large bundle.

**Figure 3.51** Sawing smaller bundles.

It may have its tubes removed from the front side using a
special puller as opposed to the tube bundle shown in
Figure 3.52, which requires that tubes be driven out from
the front side

### Driving out Tube Ends

Tubes are driven from the front of the tube sheet using a
pneumatic tool, which is a very noisy operation

The shiny part of a tube in Figure 3.52 is the part of the tube
that was rolled into the tube sheet

After cleaning of the tube sheet, tube sheet holes will be
inspected for the condition of the internal grooves, out
of roundness, and maximum diameters

If the tube sheet is within specifications, it will normally be
reused

Figure 3.52 Driving out tube ends.

## Removing Tubes from a Cage

During tube removals (Figure 3.53), tube ends are clamped and extracted from the cage using a cable winch

Because this operation is hazardous, nonessential personnel are restricted from the tube-pulling area during this phase of the disassembly

The winch operator sits behind a protective wire shield in case the winch pulling cable parts during this operation

## Assembled Cage

In Figure 3.54, a cleaned and inspected tube cage has been reassembled in preparation for the loading of new replacement tubes

A tube cage this size might take multiple 12-hour shifts to load its tubes because

**Figure 3.53** Cage being stripped of tubes prior to final disassembly, cleaning, and inspection.

**Figure 3.54** A clean cage reassembled.

There may be thousands of tubes

Each tube will need to be guided through each baffle, tube support hole, and finally into the tube sheet, which will be in place

This can be a painstaking operation, as alignment of the baffles' tube supports and finally the tube sheets have critical tolerances

### Loading the Cage

Tubes are loaded by placing alignment tube guides into the tube ends and then sliding the tubes through the cage; this is normally a two-person operation

A tube guide, as shown in Figure 3.55, is bullet shaped on one end with a short wire brush on the other end; there are other designs—this design is easy to insert and remove

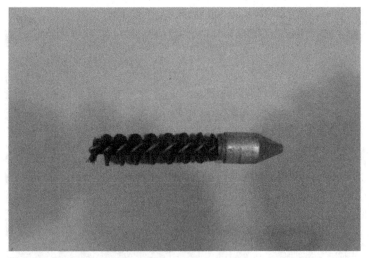

Figure 3.55 Tube guide.

**Figure 3.56** Loaded cage.

The wire brush end is pushed into the tube, and the tube is then inserted into the cage using the bullet end as a guide, which can navigate through multiple openings

In Figure 3.56, wire guides are projecting from the front and will be removed for use with the next load of tubes into the cage

Tube loading has now been completed, and tube guides are being removed in preparation for the next phase of the bundle's fabrication

There are still many hours of work ahead before this bundle of tubes will be ready for testing

Figure 3.57, a view from the side, shows more detail about the operation

Note that the front tube sheet has been mounted and that the tube ends are of uneven lengths; adjusting and cutting

**Figure 3.57** Loaded tube cage side view.

prior to rolling the tubes into the tube sheet will be accomplished after all alignment has been achieved

Some designs have a tube sheet on both ends of the tube bundle,
whereas others, such as the U-tube bundle, will have only one
tube sheet

At this stage in assembly, components of the tube cage, such
as baffles and tube supports, as well as front and back
tube sheets, are ready to be aligned and spaced to the required dimensions

### Mounting and Aligning Tube Sheets
Mounting tube sheets is done by the use of temporary nuts
and bolts

Allows the alignment and adjustment of tube sheets and
end-to-end dimension when two sheets are required

**Figure 3.58** Mounted tube sheet.

The tube support and baffle spacing are already aligned and in a fixed position

Figure 3.58 is typical of the alignment process

By tightening and loosening the nuts on the sides of the tube sheet it can be aligned with other components of the bundle

### Semiautomatic Tube-Rolling Machine

The semiautomatic tube-rolling machine in Figure 3.59

Is self-supporting

Because the operator does not have to support it during tube rolling, like a manually operated tube roller, this makes the chance of a faulty roll due to fatigue much less likely

**Figure 3.59** Semiautomatic tube-rolling machine.

Most often the semiautomatic tube-rolling machine is used for new builds

Manual tube rolling, discussed later, is essentially the same as semiautomatic

Tube rolling begins at the top of the sheet, as seen in Figure 3.60

A preliminary light roll is used to determine if the tube is being expanded into the grooves properly

The first rolled tube's ID will be measured and any required adjustment to the rolling force is made

Tubes must not be over- or underrolled and rolling must follow a sequenced pattern to prevent the tube sheet from being pulled out of alignment by uneven rolling

**Figure 3.60** Machine adjustment.

Adjustments to rolling force are performed as needed until the correct setting is found and is repeatable

A gage is inserted into the tube to measure the diameter of the tube after the first attempt at rolling; this will be repeated until it indicates the proper roll has been accomplished

The required tube's ID after rolling is based on a calculation table derived from tube material thickness and tube's OD; see Figure 3.61

It may take several attempts to arrive at the final ID required

The technician is adjusting the air pressure for the semiautomatic rolling tool, as shown in Figure 3.62

**Figure 3.61** Gaging the tube roll ID semiautomatic tube-rolling machine.

**Figure 3.62** Operator adjusts air pressure regulator to set rolling force.

This adjustment will be required after the initial rolls have
   been completed and the tube expansion into the tube hole
   has been determined using a specially designed gage
Multiple adjustments may be required in difficult rolling
   conditions, such as when
   Tubes vary in hardness
   Tube sheet holes vary slightly in diameter

## Manual Tube Rolling

Manual rolling shown in Figure 3.63 is not always as precise as semiautomatic rolling from one tube to the next as
   it is a function of how the roll feels to the technician performing the work
Manual tube rolling is often chosen for
   Retubing used exchanger

**Figure 3.63** Manual tube rolling.

Is faster than the semiautomatic, increasing production
    Used for more forgiving tube materials that do not require
        as precise and repeatable force to make a good seal at the
        tube sheet
All things being equal, the method chosen is decided by
    Availability of equipment and personnel

### Tube Rolling

If all goes right, the tube roll should match Figure 3.27
Note that the tube's internal surface has been forced into the
    tube sheet grooves by rolling of the tubes
This repeated rolling over time expands individual tube
    holes in the tube sheet—eventually to the extent that
    the tube sheet will reach a point where it may be no longer
    usable because
The tube sheet is deformed
Tube sheet holes are enlarged to the point that the tubes
    can no longer be expanded into the holes' grooves,
    assuring a leak-tight seal

### ▶ HYDROSTATIC LEAK TESTING

### Tube Bundles and Test Heads

If the complete exchanger was shipped for repairs, the chan-
    nel cover and/or the floating head can be mounted using
    temporary gaskets and then the rebuilt bundle is tested for
    roll leaks
Normally this is not the case
    The tube bundle must be tested for roll and tube leaks
        using temporary heads
    Repair facilities normally have an extensive selection of tem-
        porary tube bundle testing plates (refer to Figure 3.65) to
        accommodate most standard sizes

**Figure 3.64** Temporary heads used for pressure testing.

However, sometimes it is necessary to custom fabricate such a head, as can be seen in Figure 3.64; these will almost always be of the flat head design and will be thick

### Test Heads, Clamps, and Pressure Gages

In situ, some designs will have the tube bundle inserted into the shell with a gasket behind the bundle tube sheet and then installing the channel or bonnet if so equipped, also with a gasket used on it

As it is impractical and a last choice to ship all parts of an exchanger to the repair shop, a means of gasketing for the flat test heads of Figure 3.64 must be provided

This is accomplished using temporary clamps, referred to as homer clamps, named after the man who designed the clamps

Figure 3.65 Test clamps.

Figures 3.65 and 3.66 show homer clamps used to seal the test head and bundle tube sheet gasket during pressure testing

Note the large number of clamps needed to secure the head to the bundle in Figure 3.66 in order to make a leak-tight seal

Leaks, even small drips at the temporary head, are undesirable because

At the start of a test the test gage pressure is noted and then watched for the duration of the test to see if any unexplained pressure decay has occurred, thus indicating an unseen leak

Leaks at the gasket for the temporary head, and the resulting pressure loss on the gage, must be stopped in order to determine if there might be leak in the tube bundle elsewhere

**Figure 3.66** Rack of test clamps.

When leaks are indicated in a retubed bundle, the most
likely place for a leak is in a tube roll or at the temporary
test head gasket; leaks in older tube bundles can be located anywhere in an individual tube(s)
Such as corrosion pinholes along a tube's length

### Pressure Testing for Tube and Roll Leaks

As described previously, the test head(s) has been attached
using a temporary gasket(s) and clamps (Figure 3.67); it
can sometimes be difficult to obtain a tight seal
While under pressure the back side of the tube sheet(s) is
carefully inspected visually for any leaks at tube rolls
by looking down through the tube rows one at a time
to inspect the tube sheet between rows of tubes

**Figure 3.67** Testing for tube and roll leaks.

Laying out dry sheets of paper to look for wetting is one
 option to indicate the general area of a leak
  Often, only a small drip of water will be present
  Normally, any leak is unacceptable and the leak must be
   found; if a tube roll leak, the head will be removed and
   an attempt at rolling the leaking tube made
  With hundreds or even thousands of tubes in lengths of 20
   feet or more, a pinhole leak can be anywhere

### Assembled Exchangers

Heat exchangers usually have two sets of test pressures
 per side
  One for overload strength tests and
  One for "operating" or "leak" tests
Strength tests are
  Set by the design code and
  Shown on the original design data sheets
If original data sheets are not available, then
  Calculations must be done
  The method will depend on which design code or stan-
   dard is used
    Most common is the Tubular Exchanger Manufacturers
     Association, which uses the American Society of Me-
     chanical Engineers pressure vessel code for reference
     in this area
Some larger units will have the tube sheets specially
 designed to withstand a much lower differential pressure
 (requiring both sides to be tested simultaneously)
This information should be on the design sheets and
 on the vessel name plate (assuming that either is available)
If the intent is to check that a gasket has been installed properly, then it is permissible to perform a lower pressure test
 based on the operating pressure

The acceptability of this lower pressure test will often depend on the consequences of a leak

Two purposes for conducting these tests

Stress exchanger parts to determine structural strength

Disclose leaks through tube rolls, welds, or gaskets

Three methods for applying test pressures

**Method 1:** Pressure on the shell side with atmospheric
pressure on the tube side, then reverse this with pressure
on the tube side with atmospheric on the shell side

**Method 2:** Pressure on one side and atmospheric on
the other side side, then apply pressure to both sides
simultaneously

**Method 3:** Pressure on both sides at once (differential
test)

Which method is chosen is based on type and circumstances

If an exchanger's bundle is old and has scale and deposits
on the tubes, testing by alternating pressure on opposite
sides has the advantage of revealing small pinholes or
cracks that may be sealed by deposits such as scale on
the test side, but might be dislodged when pressurized
from the opposite side

In the aforementioned case, Method 1 might be more
appropriate to obtain the most pressure at suspected
locations

Normally the shell side would be tested first with the shell
covers off the shell for leak detection

After a successful shell-side test, the shell covers would be
installed and shell-side blinds removed for the tube-side
test

In Method 2

Pressure is applied on one side while atmospheric pressure
is on other side and then pressure is applied simultaneously to both sides

Best used for an exchanger when it is not desirable to dis-
assemble the exchanger for testing using Method 1

**Complete Exchangers**

In Method 3

Two different pressures and two gages are used, one on each side

If pressure is observed to be equalizing between the shell side and the tube side, a leak is indicated without having to remove the shell covers

This test would be required in large exchangers that have the tube sheets specially designed to withstand a much lower differential pressure, thus requiring both sides to be tested simultaneously

This test requires a longer time to determine if minute leaks exist

Consideration should be given to pressure drop when the exchanger test water might cool and the corresponding drop in pressure, which accompanies the cooling

Both a pressure gage and a temperature/pressure chart recorder may be required

Especially on larger exchangers

One concern when performing high-pressure strength tests is

Possibility of overstressing older exchangers, which may be in marginal condition and approaching end of service

This is an engineering decision and is not normally within the scope of an inspector

Full overload tests can

Approach the yield strength of some materials

Be risky if the parts have suffered wall loss due to service

If leak tightness is all that is required, such as a leak that might occur at a gasket, and consequences of a leak occurring are low, operating pressure for the test could be considered

**Example: Hydrostatic Test Pressure**

Given:

Codes:                                ASME Section VIII, Div. 1

ASME Section II, Part D

Material specification:    SA-516-Gr.70
Minimum yield strength:    38,000 psi
Tensile strength:    **70,000 psi**
Allowable stress:    (70,000/3.5 = 20,000) as per
    Post Addenda1999 Edition
Test pressure:    1.3 × MAWP

**Determine:** Hydraulic test pressure

Solution:

Effectively the 1.3 × MAWP is loading the walls at 1.3 × 20,000 = 26,000 psi during full test pressure, or about 68% of yield strength

This is based on the original minimum thickness calculated after deduction for corrosion allowance

Vessels built prior to 1999 Addenda have a Safety Factor of 4 and the allowable stress was 17,500 psi. The multiplier used was 1.5 × MAWP

## Importance of Hydrostatic Test Pressures

There are materials of construction that have much lower yield strengths than plain carbon steels

One such material is austenitic stainless steel, 316 L plate and while it has a minimum tensile strength of 70,000 psi, minimum yield strength is 25,000 psi. The first allowable stress of 16,700 psi, which is based on two-thirds of yield strength per ASME Code

Applying the same test pressure loading as our previous carbon steel example, we have 1.3 × 16,700 = 21,710 psi; this comes close to the 316 L yield of 25,000 during a hydrostatic test

Another example is SA-179 seamless cold-drawn low carbon steel heat exchanger and condenser tubes. This specification has a minimum tensile strength of 47,000 psi and a yield of 26,000 psi

**Importance of Accurate Gages**

The pressure gage is a delicate instrument and should be handled with care if an accurate indication of pressure is to be obtained

They can be inaccurate for several reasons

Rough handling

Being subjected to higher pressures than designed for

Wear and tear that comes with years of use

Gages need to be calibrated on a frequency that will ensure accurate readings are indicated during a given pressure test

If of the dial indicating (analog) design, they should be chosen for a given pressure test in the appropriate pressure range

When doing any pressure test, accurate gages are important, especially when doing overload test required by codes for new construction or alterations

ASME Section VIII Div. 1 states that dial indicating gages must

Be calibrated to assure accuracy

The desirable gage range is 2 times the calculated test pressure, but cannot be less than 1.5 times and not more than 4 times the intended test pressure

This is to keep the dial pressure indication in the center range of the gage, between 10 and 2 o'clock where these devices are the most accurate

Digital readout gages can have a greater range if proven to have the same accuracy as a calibrated dial indicating gage

Figure 3.68 shows examples of a range of gages required in any exchanger fabrication or repair facility; it shows only about 75% of the gages maintained in this shop's inventory

Figure 3.68 Assorted range of calibrated gages.

Each gage bears a calibration sticker indicating the next
    date that its calibration is due

An out-of-date or damaged gage is not acceptable

Gages are usually controlled in the tool room of a facility
    and issued upon request; the range of the gage required
    must be listed on the hydrostatic test procedure written
    and issued for the test

## ▶ HYDROSTATIC LEAK TESTING

### Importance of Water Quality

Some alloys when subjected to hydrostatic testing can be
    contaminated by the test water; this is true of the austen-
    itic stainless steel specifications of the 300 series, such as
    304, 309, 316, and others

The concern with this type of stainless steel is chloride
    stress corrosion cracking

Due to this issue, the chloride content in water used for testing these alloys should be less than 50 parts per million (ppm)

Potable water or tap water is not usually available with such a low concentration of chlorides the way to obtain suitable water is by treating the water to be used in testing

Figure 3.69 shows a set of water demineralizers, which filter and treat the water to remove chlorides and other potentially harmful chemical compounds

### ▶ BAFFLES AND TUBE SHEETS

### Tube Sheet Manufacture

When a tube sheet, baffle, or support is to be made it could be manufactured in the repair facility or sent out to a contracted machine shop

**Figure 3.69** Water treatment.

**Figure 3.70** Tube sheet blank machining.

The first step is cutting the diameter of the disc shape using the required material and machining to specification, as depicted in Figure 3.70

As with all machining, measurements are taken at several steps in the cutting operation, both by the machinist performing the cutting and by the shop floor quality control person

After the tube sheet blank has been finished, then the laying out with precision of the tube's hole sites in a blank, as shown in Figure 3.71, can be performed

In Figure 3.72, layout has been completed and hole boring has started

Alloy materials, such as grades of stainless steel, will normally require the quality control department to confirm that the chemistry of the alloy is correct by the use of positive material identification methods

**Figure 3.71** Baffle and tube sheet manufacture.

Quality control will measure the initial layout with measurements taken at several hold points during the fabrication process

### Tube Sheet and Tube Support Manufacture
The sheet stack will take considerable time to machine to its final dimension; a mistake here would be very expensive because a delay in completion of the remanufacture of this exchanger will cause
Loss of material and labor
Possible delay restarting a processing unit
Figure 3.71 shows a stack of baffles with a completely bored tube sheet on top; blank disks are underneath it
It will serve as guide for boring the blanks beneath and ensure alignment; later on the stack of baffles will be machined to the required outside diameter and cut into sections

Figure 3.72 Baffle and tube sheet stack.

## Baffle Machining

As shown in Figure 3.73

ODs of the disks are machined to design tolerances

Once machined, baffles will be cut into segments as required by baffle design

They may be quarter-moon, half-moon, or some other profile depending on the original design

This operation is performed slowly with a skilled machinist and attention to detail

## U-Tube Cage

Figure 3.74 is a bundle cage for a U-tube exchanger

This bundle has the innermost first row of tubes installed, as can be seen in Figure 3.74

**Figure 3.73** Baffle and tube sheet machining.

**Figure 3.74** U-tube cage.

As a U-tube design, it has only a front tube sheet, which represents just one of the benefits of the U-tube design, less cost

Tubes are bent to a much tighter radius in the innermost rows, and each radius gets progressively larger as the tubes radiate out from the center of the bundle

Figure 3.75 shows a completed U-tube bundle at the U bends

### Complete U-Tube Bundle

As can be seen, the flow through this design of bundle makes a 180° turn through the tubes (refer to Figure 3.76)

Some advantages of this type of exchanger are

The design is less expensive, requiring only one tube sheet

It is compact with a smaller footprint taking less space

In high temperature service it can expand and contract as required without the need of a floating head

**Figure 3.75** U-tube bundle return bends.

A disadvantage is

When in a service where hard internal deposits of product might be expected cleaning may be very difficult in the 180° bends of the tubes, especially in smaller radius inner tubes

## ▶ HEAT EXCHANGER BUNDLE REMOVAL

The bundle extractor shown in Figure 3.76 is a large powerful machine, which is stationary mounted and can remove most bundles from exchanger shells

However, sometimes a bundle is too large or is stuck in the shell of the exchanger and is too much for this machine to handle and a bigger piece of equipment is required

When all routine methods of removal have been attempted and failed, there may be no choice but to

**Figure 3.76** Shop bundle extractor.

**Figure 3.77** Large portable bundle extractor.

bring in higher tonnage equipment; one such machine is shown in Figure 3.77

Removing larger bundles from exchanger shells is best done using what is referred to as a portable bundle extractor, as shown in Figure 3.77

These machines can safely and efficiently remove even large bundles in a short amount of time

This machine is used primarily within a chemical plant or a refinery; it can be positioned and supported by a large crane and is capable of removing large bundles from exchangers installed far above ground level

Safety is a concern; note that the operator is working alone and will operate the equipment at a safe distance from the equipment, thus limiting personnel exposure (Figure 3.77)

## Bundle Removal Methods

On occasion, bundles in exchanger shells will become so fouled by product that they become extremely difficult, if not impossible, to remove by normal methods

Multiple efforts using a large bundle extraction machine may not prove successful

The bundle of Figure 3.78 could not be pulled through the shell body flange due to excessive product fouling and buildup on the tube bundle; note the hard mass on the bottom

This leads to attempting different methods to free the bundle for removal

**Figure 3.78** Product buildup on tube bundle.

## ► BUNDLE REMOVAL PROCEDURES

### Bundle Removal Attempts

Initial first attempts at removing the tube bundle in Figure 3.79 from its shell were:

Pulling on eyebolts threaded into the tube sheet the normal method used

When this failed

Removing outer tubes close to the shell using a tube ID cutter and then attempting to pull the bundle using eyelets in the tube sheet; after this failed then

Welding pulling eyes and jacking lugs on tube sheet; often the eyelet hole threads in the tube sheet will strip out during first attempts at removal, as shown in Figure 3.79

**Figure 3.79** Exchanger tube bundle fitted with pulling lugs and jacking bolts.

## ▶ TUBE BUNDLE REMOVAL

### Bundle Removal Attempts

Pulling on the lugs and screwing in jacking bolts were also
unsuccessful

When a tube bundle is so fouled by product deposits that
none of the previously discussed methods are successful,
then the last resort is to

Cut the shell along its length, as shown in Figure 3.80, and
also

Cutting around the circumference of the shell body flange
(Figure 3.81)

The shell body flange will be slid off of the bundle after it is
removed from the shell

The body flange will be reused when the exchanger is
rebuilt

**Figure 3.80** Cutting along the shell length.

Figure 3.81 Cutting off body flange (note the reference mark used during flange reassembly).

## Cutting Around the Shell
If the bundle is to be saved, then great care will be needed
  when cutting so as to not damage the tubes inside the shell
Being careful not to penetrate the shell and then grinding to
  finish the cut, arc gouging was the method used
Using this approach, the shell can be seen to open up as the
  final grinding cut is made
Note the 0° reference mark—this is for alignment of the
  body flange during the repair phase

## Arc Gouging Part Way Through the Shell
Figure 3.82 shows yet another exchanger that needed to be
  cut to allow the bundle to be removed for cleaning and
  repairs

**Figure 3.82** Arc gouging part way through a shell.

As before, arc gouging is restricted to cutting approximately 90% through the shell wall; this cut will be finished by grinding

In this exchanger, it was not necessary to remove the back head

The body flange, however, was required to be cut 360° as in the previous example shown in Figure 3.81

All of this is done in an attempt to save the bundle for reuse; in the end it may require a complete bundle rebuild, but that will not be known until the bundle is out, cleaned, and inspected.

### Finished Cut Shell

Finally, grinding to finish the longitudinal shell cut, as seen in Figure 3.83, is a slow process and must be done carefully; a mistake here would do harm to the bundle inside the shell

Figure 3.83 A cut shell.

During this particular part of the job, some concern was
  given to the possibility that the shell might spring open
  as the longitudinal cut was finished
To be on the safe side of things
  Chains were place around the girth of the shell
  Nonessential personnel were prohibited from the imme-
    diate area

### Shell and Body Flange Cut Ready for Pulling

All that remains now is moving the exchanger into position
  for the large bundle extractor to withdraw the tube bundle and tube sheet as a unit, as seen in Figure 3.84
The bundle is in place at the bundle extractor station, it has
  been secured with chains to a mounting stand and pulling
  will commence

**Figure 3.84** Shell and body flange cut and ready for pulling.

### Extraction of Tube Bundle from Shell

Now because the exchanger bundle is being pulled out
  using many tons of force, pulling is done ever so slowly
    To lessen stress on the equipment
    To provide personnel safety; a broken choker cable here
      could lead to serious injury
While the pull was being applied, two workers, not shown
  in Figure 3.85, began hammering along the length of the
  shell with sledgehammers on opposite sides of the shell
This was done in an effort to shatter some of the hard product scale that is between the shell and the tube bundle,
  causing the tube bundle to be wedged into the shell

### Debris Taken from Shell

This material (Figures 3.86 and 3.87) was removed to an
environmentally safe disposal unit

**Figure 3.85** Extracting the tube bundle.

**Figure 3.86** Debris that fell from shell.

Figure 3.87 Over five wheelbarrow loads were removed.

▶ SHELL REPAIR

**Stabilizing Tabs**

Once the tube bundle has been removed, the shell must be
    cleaned and repaired
    Internal circumference is set back to its original
        dimension
Figure 3.88 shows stabilizing tabs that were welded on the
    shell to hold the shell's circumference in place for welding
More stabilization will be needed, which will be in the form
    of strong backs
Strong backs are used to minimize the distortion that will be
    a concern during the welding processes

**Backing Bar**

Due to the large weld joint opening, the shell requires a flat
    backing bar to

Figure 3.88 Stabilizing tabs hold the shell in place.

Support the molten weld metal

Welded from the inside, which makes removal of the bar after welding the initial passes much easier

As shown in Figure 3.89, edges of the seam have been flame cut and will need to be ground to a bevel before welding can begin

## Strong Backs

Strong backs are welded to

Hold the shell in place as welding progresses

Control shrinkage as the weld metal solidifies

Strong backs may be placed on the inside or outside of the shell, depending on which side the initial weld passes will be made

Figure 3.89 Edges of the seam have been flame cut and will need to be ground to a bevel.

In Figure 3.90, manual welding of this shell's long seam has commenced

### Internal Strong Backs

As shown in Figure 3.91, quarter-moon strong backs maintain the inside circumference

In this repair, welding was performed from the outside first

After completion of the outside weld, the strong backs will be removed and the internal weld joint will be ground to smooth sound metal and the weld completed from the inside

A total of three vessels that were cut open to remove the bundles were repaired in this manner

Figure 3.90 Welding from the outside is faster.

Figure 3.91 Quarter-moon strong backs maintain the shell circumference.

## External Strong Backs

In cases where it is desirable or necessary to make the weld from the inside first

Strong backs are welded on the outside of the shell (Figure 3.92)

These strong backs were the remains of the steel plate used to make the internal strong backs shown in Figure 3.91

Because two identical vessel were being repaired, one was welded from the inside and the other outside, eliminating a waste strong back plate

Normally the shell is rolled into a nearly flat position to make welding easier

A backing bar was used as the weld prep groove has a wide spacing; the bar will be removed to complete the weld from the outside

**Figure 3.92** External strong backs.

## Flat Metal Backing Strip (Bar)

Routinely, first weld passes are performed using a manual or semiautomatic process such as shielded metal arc or gas metal arc welding (SMAW/GMAW)

The balance of the weld can then be made using the GMAW process for increased speed or SMAW as appropriate

The strong backs will be removed once the shell has had enough weld metal deposited to hold its dimension stable (Figure 3.93)

## Rabbit (Template)

After welding has been completed and the strong backs removed

Figure 3.93 Flat metal strip use as weld backing.

Any internal weld reinforcement (weld cap) is ground flush
   in exchanger shells to provide maximum clearance for the
   tube bundle
   A template known as a rabbit is passed through the shell
     (refer to Figure 3.94)
This is done to make sure the shell inside diameter provides
   for the outside diameter of the exchanger's tube bundle
   plus adequate, but not excessive, clearance

### Reattaching the Body Flange

After the shell has been welded, the shell body flange is
   fitted to the shell cylinder for welding
The shell cylinder is
   Stabilized by the completion of its longitudinal weld seam

**Figure 3.94** A rabbit (template) is pushed through the shell to check for obstructions and for dimensions.

Made round to meet ASME pressure vessel code specifications

So that the difference between maximum and minimum inside diameters at any cross section shall not exceed 1% of the nominal diameter at the cross section under consideration

The shell flange is attached using the same approach as new construction (Figure 3.95); it has to be attached using small welds in preparation for welding by the submerged arc process

### Submerged Arc Welding Process (SAW)

The body flange shown in Figure 3.96 will be attached using the SAW process

Figure 3.96 shows the process being used to weld the longitudinal weld seam of a vessel cylinder

**Figure 3.95** Shell body flange is fitted to shell cylinder for welding.

**Figure 3.96** Submerged arc welding process.

For welding the body flange shown in Figure 3.95

The exchanger shell will be mounted on an automatic rolling machine, and the SAW process is used to weld the entire circumference, employing several revolutions of the
assembly to complete

This process is relatively fast, accurate, and can have excellent control of welding heat input, thereby minimizing
distortion

## ▶ HEAT TREATMENT

### Postweld Heat Treatment (PWHT)

PWHT is one of many types of heat treatment, such as

Annealing to soften metal for forming or machining

Normalizing to provide uniformity in grain size

Quenching to harden ferrous metals

It can be required in circumstances where
  It is required by the code of construction to relieve
  the locked up stresses produced by the welding process in materials based on the thickness/material
  combinations
  In welding operations, it is can be required to relieve the
  weld metal and heat affected zone stresses created during welding for service environment reasons without regard to thickness
Locked up stress can result in stress corrosion cracking, distortion, fatigue cracking, and premature failures, as well
  as accelerated corrosion at the weld and its heat-affected
  zone
Additionally, when stress from an applied load, such as internal pressure in an exchange shell, is added to the resid-
  ual stress left by the welding process
The exchanger shell repaired in the previous shell repair section was originally postweld heat treated and was required to receive PWHT after repairs
Heat treatment can be performed by placing an entire exchanger shell in a heat-treating oven or by performing
  it locally to selected welds
Often it is more efficient to simply move the entire exchanger's vessel into an oven, as shown in Figure 3.97
  This heat treatment oven can accommodate fairly large
  and thick exchanger shells
An oven such as this is normally used when local heat treatment methods may be not be sufficient or practical
This is only one of the ovens in this exchanger repair shop;
  this shop has two, as well as local heat treatment
  capabilities
In Figure 3.98, the oven adjacent to the oven of Figure 3.97
  is loaded with an exchanger shell that is to be heat
  treated

**Figure 3.97** Medium capacity heat treatment oven.

**Figure 3.98** Exchanger shell being prepared for PWHT.

The shell has only been loaded at this stage, it still needs
   To be equipped with thermocouples for monitoring its
      temperature during the heat treatment operation
      Requires more than one thermocouple; normally a
         minimum of two are required even for small welded
         repairs
PWHT minimum temperature and time at temperature,
   referred to as soak time, are usually obtained from
   the code of construction; for other forms of heat treatment, the criterion is based on good engineering practices
Heat treatment temperatures should be controlled pre-
   cisely; Figure 3.99 is the control room for the ovens
   pictured in Figures 3.97 and 3.98
These heat treatment ovens are natural gas fired; two sets
   of controls seen in Figure 3.99 control the flow of gas

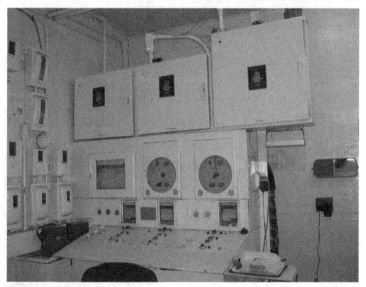

Figure 3.99 Heat treatment oven control room.

to the oven's burners, modulating the rate of firing in order to control the temperature inside the oven to a desired
preset temperature

The exchanger metal temperature is sensed by the thermocouples attached directly to the exchanger shell

Note also in Figure 3.99 that there are two chart recorders,
above the controls shown, one temperature recorder for
each oven

The chart recorder is operating the entire time during heat
treatment, and the chart becomes a permanent record and
is shipped with the heat-treated equipment to the
customer

The heat treatment process is determined by the applicable
code when done to meet Code requirements

Codes may have listed parameters, such as

Maximum temperature of the oven when the part is
placed inside the oven

Rate of increase in temperature

Required minimum holding temperature

Minimum time at temperature

Cool-down procedure

Example: Just to name a few requirements

ASME Section VIII Division 1, paragraph UCS 56 requires
that

The part cannot be placed in the oven at a temperature
greater than 800°F (425°C)

Above 800°F (425°C) the part must not be subjected to a
rate of heating greater than 400°F (222°C) per hour based
on thickness

Other requirements, such as differential of temperature
over a given length of the part, flame impingement, and
finally the rate of cooling acceptable to the code, are is-
sues that must be considered when a manufacturer or a
repair concern perform heat treatment operations

## Local PWHT

Frequently, only a portion of an exchanger part is required to be heat treated, for example, seal welds made at tube ends where leaking is occurring or as an effort to ensure that no leaks will occur during operation of newly constructed exchanger, a form of positive sealing

Figure 3.100 is a case where leaks have been experienced in an older exchange tube bundle, which requires heat treatment after weld repairs to previous seal welds at tube ends

The light stains seen on the tube sheet of Figure 3.100 are residues from previous dye penetrant examinations used to locate cracks; darker stains are flash corrosion from water leaks during pressure testing to locate leaking tube ends

**Figure 3.100** Exchanger tube seal welds.

Any additional welding after PWHT will require that the local area welded be subjected to the original PWHT requirements

The ceramic heat treatment element pad shown in Figure 3.101 was mounted using the appropriate welding filler metal with the spot welding process (SW)

Many more such pads will be attached across the face of this exchanger tube sheet, enough to ensure that the minimum temperature can be maintained across the entire surface that has been affected during seal welding of the tube ends

Figure 3.102 shows attachment of the required pads for this local heat treatment operation

In order that the heat be distributed evenly over the surface and reduce heat loss, the entire head will be covered with

**Figure 3.101** Heat treatment pad attachment method.

**Figure 3.102** Additional local heat treatment pads.

a insulating material capable of withstanding the required temperature

In this case, approximately 1300°F (705°C) ± 50°F (10°C) the actual required temperature is 1250°F (675°C), but because such precise control is difficult to maintain with such a large mass, even in a furnace, some tolerance of temperature variance is given

Finally, in Figure 3.103, the entire exchanger tube sheet has been insulated as it is ready to receive heat treatment

Monitoring during local PWHT is recorded just the same as when it is performed in a furnace

The chart is shipped to the customer and often it, along with other documentation, is handed over to the customer by the driver who makes the delivery to the plant

**Figure 3.103** Insulated and ready for PWHT.

Figures 3.104 and 3.105 show the rest of the local PWHT
  equipment
**Safety Note: Burns to personnel** must be considered; **in addition, noxious fumes** can be emitted from chemical deposits
  on the hot surface of exchanger parts—**this should always
  be considered**

## ▶ DOUBLE-PIPE EXCHANGERS

Often referred to as a hairpin due to its shape

### General Considerations

Consists of a pipe or tube inside of a pipe shell
Developed to fit applications that are too small to use
  TEMA requirements
Tubes are often finned to yield additional surface area

**Figure 3.104** Local PWHT electrical power supply.

**Figure 3.105** Local PWHT temperature/time recorder.

## Advantages
Inexpensive and readily available
Can handle thermal expansion due to U-bend construction
Has a small footprint, requiring less space

## Disadvantages
Limited to bare tube surface area of less than 500 ft$^2$ (maximum 1000 ft$^2$)
Cannot remove tubes easily
Cannot mechanically clean tube and shell easily

## Uses
Clean fluids where low heat transfer area is required with high temperature, high pressure (500 psig)

## Tube Configurations
See Figure 3.106

## Inspections
Inspections of this type of tubular exchangers are similar to inspecting shell-and-tube heat exchangers
Exchanger is disassembled, cleaned, and inspections performed
Figure 3.107 shows a disassembled stacked unit
   This is in a series configuration
   Used for steam heating a product stream, the double pipes are laying on their shells

## Visual Inspection
Figure 3.108 shows pitting on tubes of a double-pipe exchanger

**Hairpin Heat Exchanger Details**
**Double-Pipe Design**

Bare Tube

Finned Tube

Tube connection options include,
weld-end, flanged or Gray-Loc®/REFLANGE for high pressure service.

**Multi-Tube**

Bare Tube

Finned Tube

**Figure 3.106** Four common hairpin/double-pipe exchanger tube configurations.

This pitting may look severe but it is only 0.0625 in. (1.6 mm) deep with a maximum diameter of 0.375 in. (9.5 mm) in a pipe wall that was measured to be 0.231 in. (5.9 mm) thick

There were no repairs required; however, a recommendation was made to consider replacing this double-pipe assembly at the next 5-year shutdown cycle

**Figure 3.107** Bare tube double-pipe exchanger.

**Figure 3.108** Double-pipe exchanger—pitting on tube outside diameter.

## Pulling the Pipe from the Exchanger
As shown in Figure 3.109
  Inlet manifold has been removed
  Double-pipe assembly is being made ready to be pulled
  from the shell

## Hairpin Exchangers
Because these are small-diameter pipes, only a short distance
  of the pipe's ID can be inspected visually without the use of
  a bore scope
If more information is need about the ID thickness or pitting depth, then RFET could be employed
As shown in Figure 3.110, the hairpin can now be removed
  for an external and limited internal visual inspection
The cover, known as the bonnet, would also be inspected,
  along with all fittings and connections

Figure 3.109  Double-pipe exchanger prior to pulling the pipe.

**Figure 3.110** Double-pipe hairpin exchanger return bonnet removed for tube access.

### Multitube Hairpin Exchangers

Figure 3.111 shows a typical nonfinned, multitube hairpin exchanger

  Note that the tube supports (rings around the tube bundle) near the top are severely corroded

   In order to replace these tube supports, the exchanger would require retubing

   Depending on the criticality of the exchanger's service, this may not be done if the tubes are still in serviceable condition, it might be placed back in service

Refer to Figure 3.112

  Note that the body flange is split

   It has a recess for a retaining ring that must be installed to hold the bundle in place and to accommodate the gasket

  The groove seen on the circumference of the tube sheet is for a split ring retainer used to attach the manifolds

**Figure 3.111** Multitube hairpin exchanger.

**Figure 3.112** Front view of a multitube hairpin exchanger.

The inlet and outlet manifolds are not yet installed; this is one of the last steps before completion of maintenance and final acceptance of the hydrostatic leak tests

### Hairpin Tube Sheet Corrosion

Hairpin tube sheet and tube damage is no different than any other tubular exchanger bundle

As shown in Figure 3.113, note the tube end near the top of the sheet and severe thinning due to corrosive attack

Tube corrosion is progressing into the exchanger tube sheet; eventually it will reach the area where the tube roll seal is, the seal will be lost, and leaking will occur

This exchanger will require retubing to be reliable based on a requirement for a 5-year run life before the next shutdown and inspection cycle

**Figure 3.113** Hairpin tube sheet corrosion.

## ▶ INSPECTION AND REPAIR OF EXCHANGER PARTS

### Channels

Shell and tube exchangers have several different configurations with different types of heads and shells; these parts are inspected and repaired in similar ways

Figure 3.114 is a large four-pass channel in very good condition

The pass partitions are near full thickness with no mechanical or corrosion damage present, gasket surfaces are also in good condition, and bolt holes for the body flange are round as opposed to showing out of roundness damage

There are several design reasons for using multiple passes in shell and tube exchangers

One reason to increase fluid velocity in the tubes is to reduce fouling deposits

**Figure 3.114** Four-pass channel in good condition.

### Channel Cover (Dollar Plate)

Figure 3.115 shows the cover for the four-pass channel of
Figure 3.114

Grooves seen here are gasket seating surfaces that mate to
the pass partitions in the channel

The flat area just inside the bolt-hole circle is the gasket
surface for mounting to the channel

Surfaces were in very good condition and needed only
cleaning. This is not always the case, as corrosion and
mechanical damage is often found, as will be seen in
the next few examples

### Bent Partition Brace

Figure 3.116, a two-pass bonnet-style channel, shows obvious damage

The bent support bar causes the pass partition to be out of
alignment

**Figure 3.115** Cover for a four-pass channel.

Figure 3.116 A two-pass bonnet-style channel with a bent partition brace.

Making a good gasket seal at the exchanger tube sheet is
doubtful during assembly, and this brace will require
straightening to be serviceable

It is not uncommon to find this type of mechanical damage,
which is often caused by rough handling

**Weld Cracking**

The channel shown in Figure 3.117
Has cracking in a weld in the channel
Will require excavation and welding to repair properly
Cracking was found with visual inspection only
The channel was polished to reveal the extent of the
cracking, the cracked material was removed, and a weld
repair was made

Figure 3.118 is a close-up view of a weld cracking that was
shown in Figure 3.117
Again found by visual inspection

**Figure 3.117** Weld cracking in a channel.

Light grinding fully disclosed the crack

Because of its fluid service, the inspection plan called for grinding welds to examine for subsurface cracking as it is known to occur frequently in this equipment

### Dye Penetrant Testing

Figure 3.119 shows a dye penetrant test performed on a suspect body flange to channel shell weld. No defects were found during this examination.

Dye penetrant testing was appropriate since the most probable damage mechanisms in this channel's fluid service were not of a subsurface type

Dye Penetrant Testing will only reveal discontinuities open to the surface; also, magnetic particle examination is limited in this way, and they are both superficial processes, as opposed to volumetric such as radiography or ultrasonic angle beam

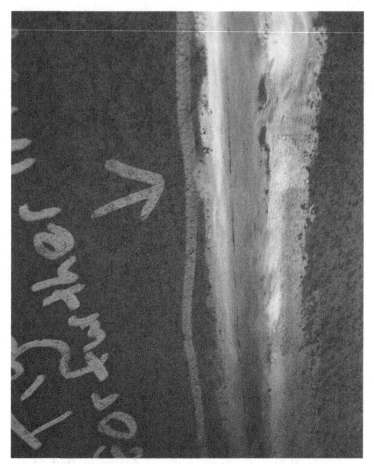

Figure 3.118 Close-up of weld cracking found by visual inspection.

## Cracked Partition Brace
A channel can have a cross brace broken near the partition attachment weld in an exchanger channel, as can be seen in Figure 3.120

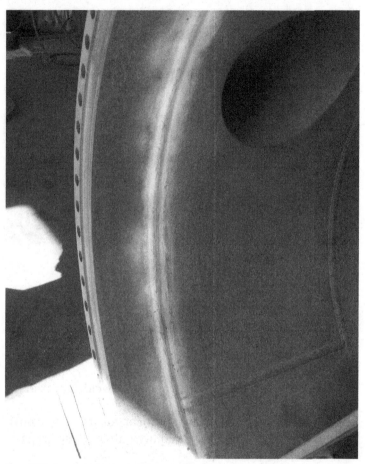

**Figure 3.119** Dye penetrant test performed on a body flange.

**Figure 3.120** Cracked partition brace.

This is a typical form of in-service mechanical damage, possibly fatigue cracking

How this break might be repaired is dependent on metallurgy, service conditions, and heat treatment condition of the channel

It might be replaced completely or simply welded back at the break

If the channel requires heat treatment for its service environment, then it would be required after welded repairs

### Damaged Gasket Surfaces

The nozzle shown in Figure 3.121 has some gasket surface mechanical damage

   The area in black is small, round, and on the outer edge of the gasket seating surface

A repair to this area would be largely cosmetic

**Figure 3.121** Damaged gasket surface.

There is no obvious leak path and repairing would be unnecessary

Of greater interest would be the weld overlay inside the nozzle bore; it should be inspected carefully for corrosion, erosion, and cracking

### Corroded–Eroded Partition Plate

As shown in Figure 3.122, the gasket edge of this pass partition, and inward for about 3 in. (75 mm), is severely corroded

Channel is in cooling water service

This would be repaired routinely by cutting out the rectangular segment marked in white.

Then welding in a small rectangular piece of plate, instead of replacing the entire pass partition, considering that the damage is limited to just this small area and the effort is to

Figure 3.122 Corroded–eroded pass partition plate gasket seating surface.

restore the gasket sealing surface on the pass partition's leading edge to ensure a good gasket sealing surface

### Pass Partition Partial Repair

The top pass partition, shown in Figure 3.123, has been repaired by replacing only the front portion of the channel partition

    This is an economical and fast turnaround repair, especially when dealing with channels used in a cooling water service

    This type of repair may not be appropriate in all fluid services due to the possibility of accelerated corrosive attack at welds

Often this type of repair can last for many years, giving dependable operation

**Figure 3.123** Partial repair of pass partition.

## Channel Shell General Corrosion
A large area of corrosive attack, near a channel inlet nozzle,
  is shown in Figure 3.124
If a repair is to be made to this area, it would be to
  Weld metal buildup followed by grinding
    Cutting out the affected area and inserting a flush plate
      patch with full penetration welding
    Placing a fillet weld patch over the area
Weld metal buildup would be the choice for most services if
  the shell is carbon steel

## Protective Coatings for Channels
As shown in Figure 3.125, this cooling water channel
  Is coated with a product called Plasticite; other competing
    products are available

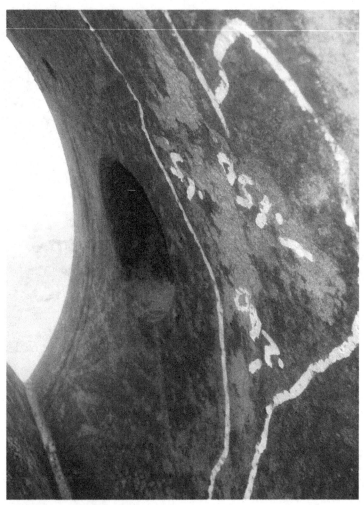

**Figure 3.124** General corrosion of a channel shell.

**Figure 3.125** Protective coating for cooling water service channel.

Used to protect from further corrosion, it is hard adhering
Surfaces must be, rust, and contaminant free for best
   adhesion to occur
When applied correctly, it can extend the life of parts

▶ **EXCHANGER ALTERATION**

**Channel Cover Alteration**

Figure 3.126 shows the first of four identical channel covers
   with a thickness of 5 in. (125 mm) that are being altered
   to equip exchanger channels with back flush nozzles
The purpose of the nozzle is to clean fouled tubes without
   removal of the channel cover
As an alteration, engineering calculations were required
   prior to commencement of the work

**Figure 3.126** Cutting a hole for a nozzle opening.

The channel cover is thick, and torch cutting by hand is difficult to make precise, especially in thicker parts
  The hole is measured and cut smaller than that required for the finished opening
  It will later be machined to its required diameter
Figures 3.127, 3.128, and 3.129 show the remaining steps in creating the rough opening in the channel cover for nozzle attachment
Figure 3.130 shows the channel cover opening of Figure 3.129 machined to final dimension in preparation for welding on the back flush nozzle
As mentioned previously, engineering calculations were done to determine if cutting this hole in the flat channel cover might cause a failure of the original design

**Figure 3.127** Checking cut.

**Figure 3.128** Slug removed.

**Figure 3.129** Rough cut.

**Figure 3.130** Machined opening in altered channel cover.

Alterations must always be investigated prior to making these types of changes per the applicable in-service code
As alterations are out of the scope of the original design and not a repair such as a replacement "in kind"
The next step is the alignment and tack welding of the new nozzle to the channel
A critical step, which is detailed in an engineering drawing
While not complicated, this must done with some precision
The channel cover is measured and marked so that there is a precise horizontal reference line at half the channel's diameter, which is then measured from to obtain the correct placement, depicted in Figure 3.131
The reference line was placed on the vertical to position the nozzle attachment so as to be accessible for ease of welding

**Figure 3.131** Channel cover nozzle placed and tack welded.

Note the two alignment pins in the nozzle flange of
Figure 3.131; these are used with a level to ensure
proper fit with the mating flange in the plant

Now the new nozzle is ready for weld out to completion,
which will be inspected visually at several steps during
welding and also by means of dye penetrant examination,
which is done routinely on the initial weld pass, also
known as the root pass

This is performed to check for shrink cracking in the root
pass, which is a relatively thin layer of weld metal; any crack
here might propagate into the following weld metal passes
and possibly into the channel cover and/or the nozzle wall

Visual inspection is performed for confirmation of proper
profile and size of the final weld

Note the level at the top of Figure 3.132; alignment of parts
is always a concern

**Figure 3.132** Alignment and leveling.

## Hydrostatic Testing after Alteration

This high-quality fit up (alignment), shown in Figure 3.133, is critical to the integrity of the welded assembly

Misalignment is one of the many causes for weld failures, as stresses can be concentrated in a poorly fabricated assembly, causing cracking and perhaps even catastrophic failure

Quality control personnel and/or inspectors will reject inferior workmanship or material knowing the danger these pose to the safety of the equipment

Any alteration to a pressure boundary involving welding normally requires that a hydrostatic test be performed

Three options were available to test nozzle welds hydrostatically

Mount the channel cover to the exchanger and test the entire tube side

**Figure 3.133** Finished fit up.

Weld pipe caps to the back side of all four channels, hydro, and cut off the caps afterward

Blind off one channel nozzle and lay it flat on stands with the inside surface facing up and then make a gasket and bolt and bolt a second cover to the first

This was the procedure chosen and is shown in Figure 3.134

## ▶ QUALITY CONTROL INSPECTIONS

### Inspection Tools

To aid in exchanger inspections, several essential tools have been developed over many years

Some such tools include

High-intensity flashlights

**Figure 3.134** Hydrostatic testing channel cover.

Measurement gages
Weld fit-up gages
Pit gages
Hammers
Tube and tube sheet hole calipers
Assortment of telescoping mirrors
Contaminate-free markers
Scrapers
Digital thickness meters
Temperature measurement devices
Positive Material Identification (PMI) Testing

### High-Intensity Flashlights

An assortment of lights (Figure 3.135) is required; larger ones put out adequate light for large exchanger shell internal inspection

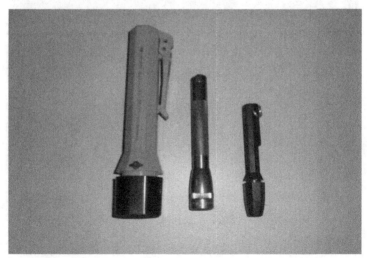

**Figure 3.135** Flashlights.

Smaller lights are usable with small diameter tubes and similar inspections of tight areas within an exchanger

Additionally, smaller flashlights can be used as backups should the larger light fail in a large dark vessel

### Weld Gages

Inspection tools are used for checking excessive Hi–Lo and if weld root spacing requirements are in accordance with welding specifications

These gages are used for checking butt weld joint fit up requirements; the upper tool in Figure 3.136 is used for checking Hi–Lo and weld root openings on welds to nozzles such as required for flanges and can be used without internal access

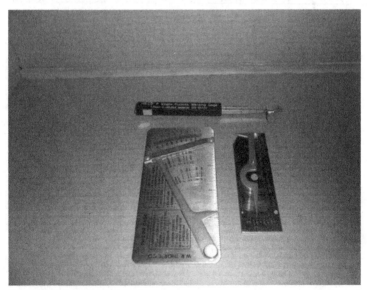

**Figure 3.136** Weld measurement gages.

The lower two gages in Figure 3.136 represent two different
   size tools that can measure Hi–Lo on external surfaces or
   an inside surface that is accessible
The most important nondestructive examination for welding inspection is visual testing because
   Many welding defects can be traced back to problems,
      such as
      Poor alignment of parts prior to welding, a condition
         known as Hi-Lo; another issue is weld root spacing
      Hi-Lo is caused by a difference in thickness between
         two parts being joined (see Figure 3.137)

### Weld Fit Up Measurements

The Hi-Lo seen in Figure 3.137 may not be obvious from
   the outside of a nozzle fit up; however, it can be found easily using the Hi-Lo inspection tool shown in Figure 3.138
When there is too much or too little opening at the root of
   the weld, it can cause excessive weld metal penetration in
   the first case or inadequate penetration in the second
Such conditions are improper weld joint preparation—a
   typical case of this is when surfaces of the weld joint bevel
   have been left in the flame-cut condition, leaving ridges
   and valleys; this condition can result in a lack of fusion
   between the base metal and the deposited weld metal
The tools shown in Figure 3.138 include the weld fit up
   tools needed to check for Hi-Lo and root opening refer-
   enced in Figure 3.137; however, these gages also serve a
   second purpose

**Figure 3.137** Nozzle wall fit up.

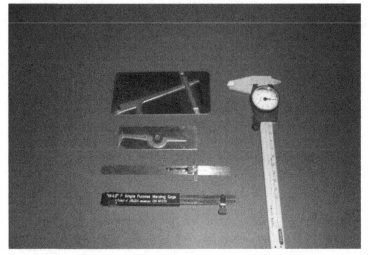

**Figure 3.138** Inspection tools.

The two top gages shown in Figure 3.138 can also be used for measuring the depth of corrosion pitting or a larger corroded area

The small ruler is often used to measure diameters and serves as a point of reference in photographs to indicate the size of a defect

Calipers are often used for measuring the ID and OD of tubes and other components

Figures 3.139 and 3.140 demonstrate use of the larger gage of Figure 3.138 and a ruler for a photograph to be sent to the engineering department of the owner–user for evaluation

If photographs like these clearly indicate dimensions, then a more competent evaluation will be possible without the engineers having to make a visit to the repair facility to make an evaluation

**Figure 3.139** Depth of erosion.

**Figure 3.140** Width of erosion.

## Hammers

Figure 3.141 shows two types of hammers; the small light ball peen hammer is used to "sound metal"

By progressing down the length of a vessel or exchanger shell and striking the wall lightly, an experienced individual listening for a metallic ring can recognize a change of sound to a duller thud as a locally thinned area of the metal

This method is an expedient way to find areas that need further investigation using an electronic ultrasonic thickness meter

**Safety: Using the hammer test on vessels under pressure must not be done; it is strictly a depressurized method of thickness inspection**

## Telescoping Mirrors, Magnets, and Magnification

Mirrors (Figure 3.142) provide access for visual inspection in tight or otherwise inaccessible areas; an assortment of sizes is desirable

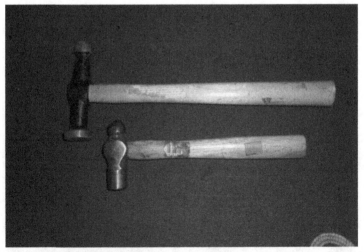

Figure 3.141 Inspection hammers.

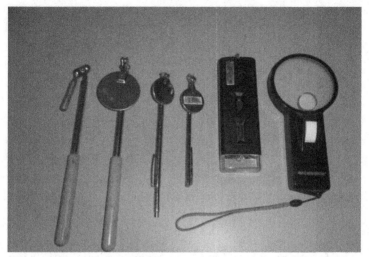

**Figure 3.142** Mirrors, magnet, and magnifier.

Magnets (Figure 3.142) can be used to do a gross screening of base metals that might not be plain carbon steel, such as the 300 series of stainless steels

Magnification (Figure 3.142) may be used to determine if a linear indication is a scratch or crack

### Contaminate-Free Markers and Scrapers

From top to bottom (Figure 3.143), the first marker shown is one used commonly by welders for layout lines prior to cutting metal, it is the soapstone; it may not be suitable for marking all metals

The second and third markers are soft and will write on most surfaces, even when corroded or uneven but smear very easily

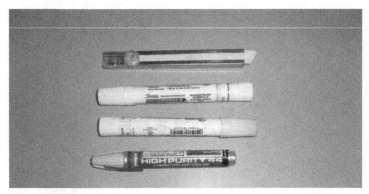

**Figure 3.143** Markers.

The fourth marker is for use on nonreactive metals; it is a
  paint for metals and is equipped with a felt tip, which
  can be rewetted should the cap be left off

### Digital Tube and Tube Sheet Hole Calipers

The gages shown in Figure 3.144 are used for checking the
  inside diameter of holes drilled in an exchanger tube
  sheet

This gage can also be used for measuring the out of roundness of holes in tube sheets, which are to be reused during the retubing of an exchanger bundle; the tool is
  rotated in the hole and measures the diameter at multiple
  places

When tolerances for a near perfect circle are not met, then
  tubes rolled into such a tube sheet hole will not make a
  leak-free seal

Note the attachments in the storage box (Figure 3.144);
  these are mounted to the meter as needed to measure
  holes by size range

**Figure 3.144** Gages and attachments.

### Digital Thickness and Positive Material Identification Meters
The upper device in Figure 3.145 is a hand-held nuclear
analyzer used to determine the content or chemistry of
alloyed metals; this model can report alloys by percentage
content or by a match to a known alloy specification such
as 304 stainless steel

The lower meter shown in Figure 3.145 is an ultrasonic
thickness meter—it utilizes ultrasound to measure the
thickness of the metal; this meter must be calibrated on
known thickness standardized blocks made of the same
or similar metal just prior to use

### Temperature Measurement Devices
Welding procedures must list requirements for preheating
the metal immediately prior to commencing welding, and
preheat must maintain ahead of the weld puddle in all directions for a given distance as specified by the applicable code

**Figure 3.145** PMI and digital thickness measurement.

Prior to technology advances and still commonly used to-day is the temperature indicating stick, made of a material that will melt when the temperature marked on a stick is met or exceeded

Advances in technology today offer the convenience of being point-and-shoot devices; giving instantaneous temperature by pushing a button (Figure 3.146)

### Individual Tube Leak Air Test

Using the device shown in Figure 3.147, individual tubes can be tested for through wall leaks such as pitting holes or cracks

Air pressure is introduced at one end of the tube while the other end is closed off with a rubber stopper and then the pressure gage is observed for pressure loss

Individual tube leak tests can take an appreciable amount of time, especially when the exchanger has a large number of tubes

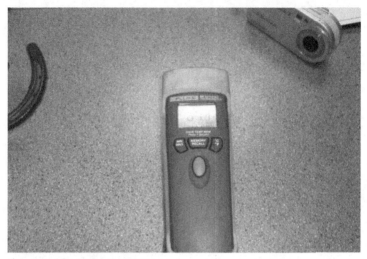

**Figure 3.146** Infrared temperature indicator.

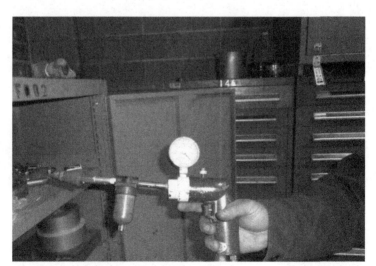

**Figure 3.147** Tube leak air tester.

# Repair, Alteration, and Rerating

**4**

## ▶ OVERVIEW

This section discusses
  Repair, alteration, and rerating of pressure vessels
  Pertinent code and jurisdiction requirements
  Differences among
    ASME Code
    National Board (NB) Inspection Code
    API 510

### Topics Include
Design of repairs
Planning and approval
Responsible organization
Materials
Replacement parts
Postweld heat treatment (PWHT)
Inspections and hydrotest procedures
Documentation and name plates

### General Considerations
Inspection of a pressure vessel frequently reveals that some
  form of deterioration has occurred during service
Analysis of the deterioration may indicate that the vessel is
  **Repaired** under the original design conditions
  **Rerated** for less severe design conditions
Rerating for new design conditions may also be necessary
  because of **changes** in **operating requirements**

New process requirements may also require alterations to a
vessel

The size and location of a vessel in a facility/plant can make
repair or alteration difficult

Nondestructive examination (NDE) of repairs and alterations is very important to assure that high integrity
has been obtained

More extensive NDE than was required for original construction is usually advisable

## ▶ CODE AND JURISDICTION REQUIREMENTS

### ASME Code

Applies directly only to the "design, fabrication and inspection during (the original) construction of pressure
vessels"

Applicability terminates when an authorized inspector au-
thorizes application of the code stamp

The ASME Code should not be interpreted to imply that all
design details and fabrication procedures not covered by
its rules are unsatisfactory for repairs and alterations

The ASME Code is formulated around design details
and fabrication procedures obtainable with good shop
practices

### Jurisdictional Requirements

Most authorities having jurisdiction require owner–operator to operate and maintain their vessels in a safe
condition

The majority of these authorities have established regula-
tions that refer to either

National Board Inspection Code

API Pressure Vessel Inspection Code (API 510) for the repair, alteration, and rerating of pressure vessels

Most companies prefer to use API 510 because it is specifically oriented to the needs of the hydrocarbon processing industry

Some authorities having jurisdiction make the owner–operator responsible for obtaining approvals and filing the documentation for repairs, alterations, and rerating

## National Board Inspection Code vs. API 510

Technical requirements are **similar**

Procedural and administrative aspects **differ**

Major differences are that the **National Board Code**

> Requires an **authorized inspector** to hold a **commission** from the National Board
>
> **Restricts** the **authority** of an **authorized inspector** employed by the owner–operator
>
> Requires **preparation and approval of an R-1 form** and attachment of a new name plate for repairs and alterations that do not change the maximum allowable working pressure and/or design temperature

The more elaborate procedural and administrative details of the National Board do not result in repairs and alterations with higher integrity, but they can increase the costs incurred considerably

Figure 4.1 is a flowchart that compares the major requirements of API 510 to those of the National Board

## API 510

Permits **greater flexibility** through the exercise of **engineering, judgment** by the owner–operator than is possible when following the National Board

Owner–operator has **more responsibility** for the integrity of a repair

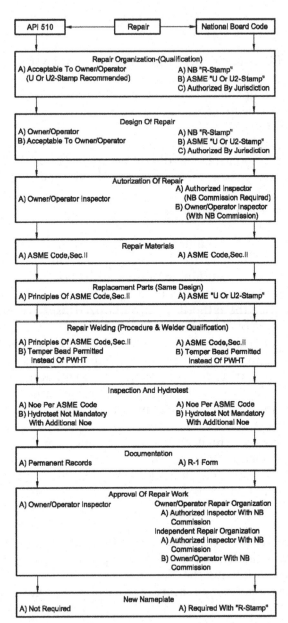

* Deviations Permitted Within Good Engineering Practice That Is Responsibility Of Owner/Operator

**Figure 4.1** Flow for repair: summary of requirements of API 510 vs. National Board Inspection Code.

### References to ASME Code

Both refer to the ASME Code for making repairs, alterations, and reratings of pressure vessels

Wording used by the National Board and API convey different implications

**National Board**

Requires all repairs and alterations to **conform to** the ASME Code whenever possible (Paragraph R-100)

**API 510**

Requires following the principles of the ASME Code

Both codes recognize that it may not always be possible to adhere strictly to the ASME Code

National Board implies that the code must be complied with whenever possible

API permits more flexibility for deviation from the code by exercising engineering judgment

Design details for repairs and alterations that deviate from the rules of the code should be justified by a stress analysis to verify that the maximum allowable stress permitted by the code is not exceeded

Fabrication procedures that differ from the original construction must be qualified properly to verify that

Minimum materials properties (strength and $C_v$ impact toughness) required by the code are obtained

Any other materials requirements specified for the service conditions (such as maximum hardness of the weld metal and heat-affected zones) are achieved

### Authorizations and Approvals

Both NB and API require obtaining authorizations from an "authorized inspector"

NB requires that inspectors **hold a commission** from the NB (Chapter 1)

API requires only that the inspector **be qualified** to perform the inspection (Paragraph 4.2.4) by virtue of his knowledge and experience (many company inspectors are commissioned by the NB)

NB emphasizes compliance with its rules

Through scrutiny of authorized inspector

Consistent with its dictum to conform to the ASME Code

API emphasizes compliance

Relies on pressure vessel/materials engineers to assure integrity

Allows the authorized inspector to base his authorization and approvals on consultations with the pressure vessel/materials engineer

API's practice follows from its underlying concept of adhering to the principles of the ASME Code while allowing flexibility to use engineering judgment

Both codes permit the authorized inspector to be an employee of the owner–operator, but the NB prohibits an employee from approving work performed by his employer unless the authority having jurisdiction (or NB) has given its consent upon review of the owner–operator inspection procedures (Paragraph 301.2d). The API contains no such restriction

### Reports, Records, and Name Plates

NB establishes a **formal administrative procedure** for documenting and recording repairs, alterations, and reratings of pressure vessels

An "R-1" form, shown in Figure 4.2, must be completed by the organization performing the work (with the exception of routine repair) and submitted to an authorized inspector for approval

FORM R-1, REPORT OF WELDED OR REPAIR ALTERATION as required by the provisions of the National Board Inspection Code

1. Work performed by_____  _____
              (name of repair or alteration organization)              (P.O. no., job no., etc)

              (address)
2. Owner_____
      (name)

              (address)
3. Location of installation_____
             (name)

              (address)
4. Unit identification:_____ Name of original manufacturer:_____
             (boiler, pressure vessels)

5. Identifying nos.: _____  _____  _____  _____  _____
         (mfr's. serial no.) (original National Board no.) (Juridiction no.) (other)  (year built)
6. Description of work: _____
         (use back, separate sheet, or sketch of necessary)

_____
_____
_____
_____

7. Replacement Parts. Attached are Manufacturers' Partial Data Reports properly identified and signed by Authorized Inspectors for the following items of this report:

_____
_____
_____
_____
_____
_____
_____
_____

        (name of part, item number, mfg's. name and Identifying stamp)
8. Remarks:_____
_____
_____
_____
_____
_____

This form may be obtained from The National Board of Boiler and Pressure Vessel Inspectors, 1055 Crupper Ave., Columbus, OH 43229
NB-66
Rev. 5

**Figure 4.2** R-1 Form Prescribed by the National Board Inspection Code for repair and alteration of a pressure vessel.

## DESIGN CERTIFICATION

The undersigned certifies that the statements made in this report are correct and that the design changes described in this report conform to the requirements of the National Board Inspection Code.

ASME Certificate of Authorization no._____ to use the _____symbol expires _____, 20____

Date _____, 19___ _____signed _____

(name of organization)                                     (authorized representative)

## CERTIFICATE OF REVIEW OF DESIGN CHANGE

The undersigned, holding a valid Commission issued by The National Board of Boiler and Pressure Vessel Inspectors and certificate of competency issued by the state or province of _____ and employed by _____ _____of _____ has examined the design change as described in this report and verifies that to the best of his knowledge and belief such change complies with the applicable requirements of the National Board Inspection Code. By signing this certificate, neither the undersigned nor his employer makes any warranty, expressed or implied, concerning the work described in this report. Furthermore, neither the undersigned nor my employer shall be liable in any manner for any personal injury, property damage or loss of any kind arising from or connected with this inspection, except such liability as may be provided in a policy of insurance which the undersigned's insurance company may issue upon said object and then only in accordance with the terms of said policy.

Date _____, 19___ Signed _____ Commissions _____

(Authorized Inspector)   (National Board (incl. endorsements), state, prov., and no.)

## CONSTRUCTION CERTIFICATE

The undersigned certifies that the statements made in this report are correct and that all construction and workmanship on this _____ conform to the National Board Inspection Code.

(repair or alteration)

Certificate of Authorization no. _____ to use the _____ symbol expires _____, 20 _____

Date _____, 19 ____ _____ Signed _____

(repair or alteration organization)          (authorized representative)

## CERTIFICATE OF INSPECTION

The undersigned, holding a valid Commission issued by The National Board of Boiler and Pressure Vessel Inspectors and certificate of competency issued by the state or province of

_____ and employed by _____ of _____ has inspected the work described in this report on _____ 19 ____ and state that to the best of my knowledge and belief this work has been done in accordance with the National Board Inspection Code. By signing this certificate, neither the undersigned nor my employer makes any warranty, expressed or implied, concerning the work described in this report. Furthermore, neither the undersigned nor my employer shall be liable in any manner for any personal injury, property damage or loss of any kind arising from or connected with this inspection, except such liability as may be provided in a policy of insurance which the undersigned's insurance company may issue upon said object and then only in accordance with the terms of said policy.

**Figure 4.2** *Cont'd*

Copies of the "R-1" form are sent to the owner–user, authority having jurisdiction, and national board (alterations only) for a permanent record

API requires only that the owner–user maintains permanent records that document the work was performed

**Both organizations** require attaching a **new name plate** adjacent to the original name plate when a vessel is **altered or rerated** (refer to Figure 4.3 for an example)

NB also **requires** attaching a **new name plate** to a vessel that has been **repaired** (with the exception of routine repairs), as shown in Figure 4.4; API has no requirement

## ▶ REPAIRS

### General Considerations

**Repair** of a pressure vessel is the work necessary to restore the vessel to a suitable condition for safe operation at the **original design pressure and temperature,** providing that there is no change in design that affects the rating of the vessel

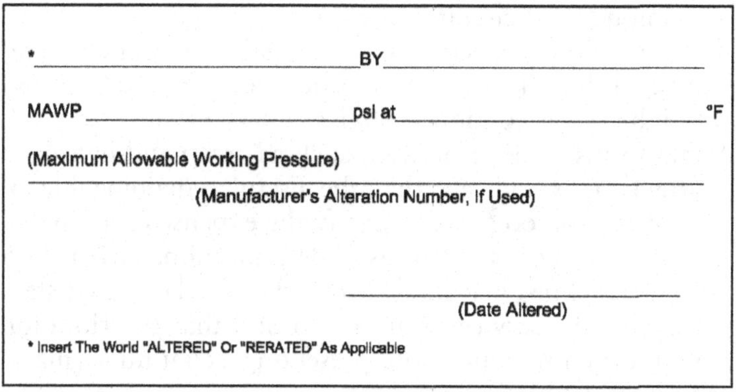

```
*_____BY_____

MAWP _____psi at_____°F

(Maximum Allowable Working Pressure)
_____
          (Manufacturer's Alteration Number, If Used)



                              _____
                                    (Date Altered)
* Insert The World "ALTERED" Or "RERATED" As Applicable
```

**Figure 4.3** New name plate prescribed by national board inspection code for a pressure vessel that is altered or repaired.

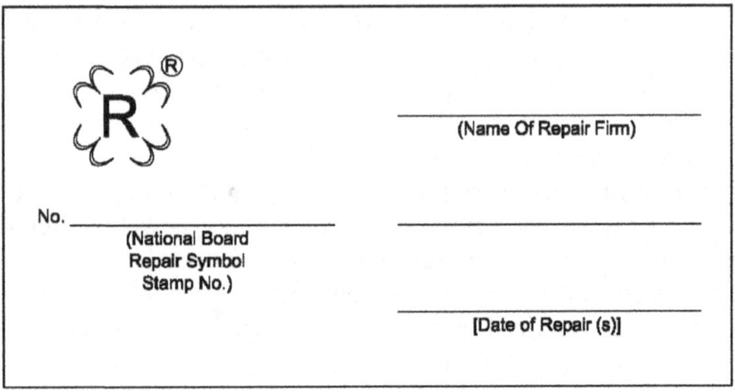

**Figure 4.4** New name plate prescribed by National Board Inspection Code for a pressure vessel that is repaired.

When deterioration renders a vessel unsatisfactory for continued service, it must either be repaired or be replaced

Figure 4.5 provides a "decision tree" that can be used to decide between repair and replacement

    Major factors that should be considered in making the decision are shown

    It may be necessary to deviate from these steps because of unique local circumstances

It is generally more economical to repair a vessel than to replace it, but the primary considerations are integrity and reliability for continued service

Some forms of deterioration, such as creep and hydrogen attack, may indicate that the useful remaining life of the vessel is too short to justify the expense of a repair

The detection of other forms of deterioration, such as $H_2S$ stress cracking, may indicate that the vessel is not satisfactory for the service environment and that deterioration will recur after repair, thus presenting a continuous maintenance problem

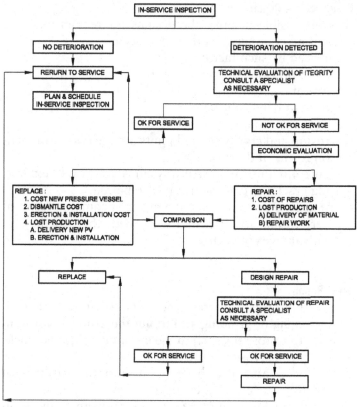

**Figure 4.5** Logic tree for repair or replacement of a pressure vessel.

## Design of the Repair

It is impossible to prescribe designs for repairs that will cover all contingencies that may occur

Influenced by such factors as

Design pressure and temperature

Form of deterioration

Extent of deterioration

Process environment

**Corrosion** and **cracks** at weld joints are the two most common forms of deterioration

Figure 4.6 illustrates three ways in which vessels exhibiting cracks or corrosion can be repaired

Some general approaches to the repair of common forms of deterioration are discussed next, with the benefits and disadvantages of each

## Weld Repair

The simplest repair of cracks consists of removing a crack by **gouging or grinding** and **filling the groove with weld metal** to restore the shell to the minimum required thickness, plus corrosion allowance

A corroded area can be ground smooth and free of corrosion scale and then restored to the minimum required thickness plus corrosion allowance by weld buildup

The ground area should be examined by MT or PT to be certain that all of the cracked or otherwise damaged material has been removed

Welding procedures should be similar to those used for original construction of the vessel, including preheat and postweld heat treatment and must be qualified according to the ASME Code, Section IX

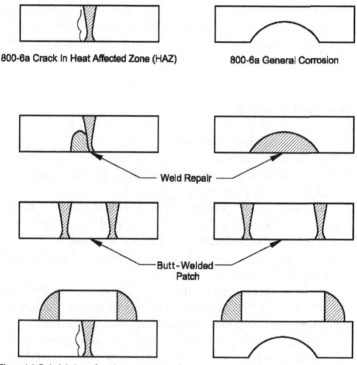

800-6a Crack In Heat Affected Zone (HAZ)

800-6a General Corrosion

Weld Repair

Butt-Welded Patch

**Figure 4.6** Typical designs of repairs to pressure vessels.

The entire weld repair should be examined by UT or RT for internal flaws and by MT or PT for surface flaws to assure that the repair will have adequate integrity for continued service

### Insert Patch

When the repair covers a relatively large area, it may be more economical to repair cracks and corroded areas by removing the deteriorated area and replacing it with a butt-welded insert patch

If the deterioration area is very large, it may be beneficial to replace the entire component of the vessel. There is usually no difference between a weld repair and a butt-welded patch with regard to integrity and reliability

### Insert Patches (Flush Patch)

Patch should be made from the same material specification and grade as the vessel components

If material is obsolete, the patch should be made from a material that has similar strength, $C_v$ impact toughness, and welding characteristics

Procurement of the proper material with the correct thickness can add to the "lead time" for making a repair and should be planned as far in advance as possible

Should be formed to the radius of the shell component that will be used to repair

Square or rectangular patches should have rounded corners with a radius of at least four times the thickness

Care should be taken to obtain as good as possible fit up of the butt patch with the actual size and shape of the opening in the shell into which they will be inserted; this fill up should be maintained during welding through the use of temporary attachments (clips) to the patch and shell

Hot cracking of the weld metal on cooling

Can be a problem with butt patches due to the relatively high restraint imposed by the surrounding vessel shell

Preheating can help alleviate this problem

Distortion (flattening) of the patch can also occur

Shrinkage of the weld metal upon cooling tends to pull the patch flat with respect to the matching radius of the vessel shell, and this can be of concern because it will change the stressor developed in the patch

The use of "strong backs" attached temporarily to the patch can minimize distortion

Every effort should be made to maintain the patch and adjoining vessel shell within the ASME Code tolerances for out of roundness to make certain that stresses developed in the patch do not exceed those permitted by the code

All clips and strong backs should be removed after butt welding of the patch to the shell is completed by cutting above the attachment welds

After their removal, the locations where they were attached to the patch and adjoining shell should be ground flush and smooth

### NDE Examination

The butt weld should receive full RT or UT examinations (regardless of the examination performed during construction) to assure the integrity of the repair

**Full coverage examination** is important because of the **possibility of hot cracking** of the weld metal

MT and PT examinations should be made on all locations where the temporary attachments (clips and strong backs, etc.) were removed from the patch and adjoining vessel shell

### Fillet-Welded Patches

Sometimes thought of as an expedient way to avoid encountering lengthy out-of-service periods

Does not conform to the definition of a repair because it constitutes a change in the design of a pressure-containing component

Thus, should be handled as an alteration

API 510, Paragraph 8.1.5.1.2 recognizes the utility of this type of repair and provides guidelines for its design

## API 510 Requirements

A fillet-welded patch should provide equivalent integrity and reliability as a reinforced opening

The primary membrane stress in the lap patch should not exceed the maximum allowance design stress for the material given in the ASME Code, and the elastic strain in the patch should not result in fillet welds being stressed above their maximum allowable stress

## Joint Efficiencies

A major constraint in designing a lap patch is to meet the joint efficiency permitted by ASME Code for lap welds (Table UW-12)

Because it is not possible to examine adequately by RT or UT the fillet welds for a lap patch to assure the absence of flaws, the efficiency of a lap-welded joint is limited to 0.50

Because of low joint efficiency, it is almost impossible to design a lap patch that will provide equivalent integrity as the original design and conduction

**Fillet-welded patches are therefore not recommended for making permanent repairs**

## Planning and Approval

Both the NB (Paragraph R-301.1) and the API 510 (Paragraph 5.1.1) require obtaining authorization for making a repair from an authorized inspector before work is initiated, except for routine repairs when prior approval has been given by an authorized inspector

Examples of routine repairs are:

   Weld buildup of corroded areas

   Application of corrosion-resistant weld overlay

   Addition of nonpressure-containing attachments when PWHT is not required

Replacement of flanges

Authorization for making a repair that is not routine is obtained from an authorized inspector by preparing and submitting a repair plan

The repair plan should be prepared by an engineer at the facility/plant in consultation with a maintenance coordinator

The repair plan should include the following information
Areas of vessel to be repaired
Repair procedures to be used for each area specifying
Preparation for repair (removal of deterioration)
Materials
Welding procedures
NDE of repairs

Repairs that will be made by a contractor should be **discussed with the contractor** to obtain his **agreement** with the plan before it is submitted to the authorized inspector

Under "emergency repairs," the repair can be initiated prior to submitting the plan to the authorized inspector, but complete documentation should be preserved and submitted to the inspector for his acceptance as soon as possible. The vessel cannot be returned to service unless acceptance of the repair has been obtained from the authorized inspector

### Organization Making Repair

The NB (Paragraph R-404) requires the organization performing a repair to have either a certificate of authorization (COA) from the NB for use of an "R" stamp or a COA from ASME for use of a "U" stamp

API 510 (Paragraph 1.2.13) also accepts an organization having an ASME "U" stamp as qualified to make repairs, but makes no mention of a NB "R" stamp

API 510 permits owner/operators to repair their own vessels in accordance with its requirements and to have repairs made by contractors whose qualifications are acceptable to them

All repairs that are not routine should be performed by an organization that has a valid "U" stamp, regardless of code or jurisdiction requirements that might permit repair by other organizations

### Repair of Materials

Both NB (Paragraph R-305) and API 510 (Paragraph 5.2.7) require that materials used for repair must conform to one of the specifications in ASME Code, Section II

Materials should be the same as those used for original construction. If this is not possible, alternate material selection should be discussed with materials engineers

### Replacement Parts

A repair can involve replacing a deteriorated part with a new part of the same design that is manufactured in a shop

Manufacturing a replacement part generally requires welding

If the ASME Code requires inspection of the weld joints by an authorized inspector, the NB requires the replacement part to be manufactured by an organization that has an ASME certificate for a "U" stamp (Paragraph R-307.1c)

A "U" stamp with the word "part" is applied to the part when it is accepted by an authorized inspector

Replacement parts do not require inspection by an authorized inspector and are not required to be manufactured by a holder of an ASME COA (Paragraph R-307.1b)

API 510 requires replacement parts to be manufactured according to the principles of the ASME Code, but has no requirement concerning qualifications of the manufacturer (Paragraph 8.1.3)

It is recommended that all replacement parts be manufactured by an organization that has a COA from ASME for use of a "U" stamp

### Repair Welding

NB (Paragraphs R-302.1 and R-302.2) requires

Qualification of all welding procedures **according to** ASME Code

Welders to pass a welder performance qualification for each welding procedure used

Repair organization to make the records of procedure and performance qualifications available to an "authorized inspector" before actual repair welding begins

API 510 (Paragraph 8.1.6.2) requires

Repair organization to qualify all welding procedures and welders used for a repair **according to the principles** of ASME Code, Section IX

"... according to principles" allows more flexibility for deviating from a welding procedure acceptable to the ASME Code when necessary to expedite a repair

Welding procedures should not deviate from the code unless reviewed by a materials engineer

### Postweld Heat Treatment

Can be a very difficult aspect of the repair and, when performed improperly, can cause additional damage to the vessel

Repair welds should receive the same PWHT used for original construction whenever possible

PWHT of a repair weld is especially important when it was specified for the original construction of the vessel to prevent stress corrosion cracking

PWHT of a repair weld is accomplished most often by local application of heat to the repaired area while the remainder of the vessel is at ambient temperature

During PWHT, high thermal stresses that can damage the vessel may develop because of severe temperature gradients and restricted thermal expansion

Nozzles, head-to-shell weld joints, attachment welds for vessel supports, piping connections, and internal components are particularly vulnerable to damage

The vessel must be free to expand when the local area is heated, and efforts should be made to keep temperature gradients less than 100°F per foot along the surface and 100°F per inch through the thickness at temperatures above 400°F

When it is not possible to perform a local PWHT within these guidelines, the risk of damage to the vessel should be evaluated carefully and alternatives for repair without PWHT should be considered

Both NB (Paragraph R-303.2.2) and API (Paragraph 8.1.6.4.2) permit substituting a temper bead (or half bead) welding procedure for PWHT for the repair of carbon steel vessels

Neither organization requires a separate qualification of this welding procedure to demonstrate that the weld metal and heat-affected zones of the repaired vessel will have the properties required to assure adequate integrity for continued service (minimum strength, maximum hardness, and $C_v$ impact toughness)

ASME Code, Section IX contains superior requirements for qualifying and performing this type of repair weld

These procedures should not be used for the repair of a vessel unless discussed with a materials engineer

**Inspection and Hydrotest**

Inspection of repairs

Both NB (Paragraph R-301.2) and API 510 (Paragraph 5.8.1.1) require the acceptance of repairs to a pressure vessel by an authorized inspector before the vessel is returned to service

The authorized inspector will normally require performing all NDE examinations for the repair that was required by the ASME Code during original construction

Alternative NDE methods can be proposed (such as substitution of UT for RT) when it is not possible, or practical, to use the NDE method used during construction

All repair welds to vessels should be subjected to essentially full-coverage NDE in view of the more difficult working conditions usually encountered for repairs compared to favorable conditions in a fabrication shop

UT is an entirely acceptable NDE method for verifying the quality of welds and does not involve the hazards and obstruction of other work associated with RT

Hydrotest after repairs

Neither NB (Paragraph R-308.1) nor API 510 (Paragraph 5.8.1.1) makes it mandatory to perform a hydrostatic pressure test following the repair of a pressure vessel, but agreement from the authorized inspector is required for it to be waived

The purpose of the hydrotest in the ASME Code is to **detect** gross **errors in the design** or **major flaws in the construction** of a new vessel

Repair of a vessel restores it to a satisfactory condition without any change in design and, therefore, there is no need to verify the design of the repaired vessel

Full coverage NDE of all repairs will detect much smaller flaws than those that could cause failure during a hydrotest and will, therefore, provide a greater assurance of the quality of the repair than a hydrotest

In-service inspection after repair

Should be planned and scheduled after it has been returned to service to assure that the repair is providing sufficient integrity

Especially important when the repair has been by deviating from some of the details of the original construction or from the ASME Code rules

NDE methods should be used to detect deterioration of the vessel that necessitated the repair

### Approval of Repairs, Documentation, and Name Plate

NB (Paragraph R-402) requires

Repair organization to document the repair by completing the R-1 form submitted to the authorized inspector for approval (Figure 4.2)

Subsequent to obtaining approval of the R-1 form, the repair organization must attach a new name plate

Name plate is stamped with an "R" if the repair organization has a COA from the NB

Repair organization cannot stamp the name plate with a "U," despite using its COA from the ASME to qualify it for making the repair

ASME only permits the "U" stamp for the design and construction of a new vessel

Completion of an R-1 form and attachment of a new name plate may not be required for routine repairs, dependent upon consent of the

Authorities having jurisdiction

Approval of the authorized inspector

API 510 (Paragraph 8.1.2.1)

Requires documentation of repairs to be kept as permanent records, but does not prescribe using a standard form

Does not require attaching a new name plate to a repaired vessel

A new name plate should not be attached to a vessel after a repair unless the authorities having jurisdiction mandate following the NB

The original name plate of the vessel has the primary purpose of permanently displaying
MAWP
Temperature rating

A repair does not change the rating of the vessel, and thus, a new name plate is unnecessary, unless required by the authorities having jurisdiction

### ▶ ALTERATION

#### General Considerations

Alteration of a pressure vessel

Physical change to any component that affects pressure-containing capability

Can change the MAWP and temperature rating of a vessel from that given on the original name plate with a "U" stamp

Alterations can be designed so as not to affect the original rating of a vessel, provided the operating pressure and temperature are not changed

Alterations are usually made to accommodate changes in process design

Installation of new nozzles

Changes in internal components

The effect that the design loads on the new internals (pressure drop, static weight, liquid head, etc.) have upon the stressor in the vessel shell should be calculated to determine if the MAWP of the vessel has to be changed

## Planning and Approval
NB (Paragraph R-501) requires **all alterations to conform to** the ASME Code

API 510 (Paragraph 8.1.1) requires **adhering to the principles** of the ASME Code
> Wording allows more flexibility for designing alterations when it is not advisable or practical to conform to the code

Both the NB (Paragraph R-301.1.2) and API 510 (Paragraph 8.1.2.1) require authorization from the authorized inspector prior to initiating an alteration

Authorized inspector will
> **Verify** that the design of the **alterations** and calculations have followed **ASME Code criteria**
>
> **Determine** that **acceptable materials** will be **used**
>
> **Assure** that the **weld procedures and welders are properly qualified**

API 510 requires the inspector to consult with an experienced materials engineer before giving authorization to proceed with the alteration

## Organization Making Alterations
NB (Paragraph R-505) requires that an organization performing an alteration has an ASME COA covering the scope of work involved

API 510 does not contain specific requirements for an organization performing an alteration
> Presumably, the same requirements would apply as for a repair organization

Alterations can be designed by qualified pressure vessel and materials engineers, but it is recommended that only

organizations holding an ASME COA perform work
on the vessel

## Materials, Replacement Parts, Welding, PWHT, and Inspection
Requirements for the alteration of pressure vessels concerning materials, replacement parts, welding, PWHT, and inspection are identical to those for repair

## Hydrotest after Alterations
Hydrotesting alterations is a mandatory requirement of the
NB (Paragraph R-308.2)
API 510 (Paragraph 5.8.1.1) states that
Hydrotesting is normally required after an alteration
Permits waiving the hydrotest after consultation with a
materials engineer if superior designs, materials, fabrication procedures, and inspections are used
A hydrotest should be performed after an alteration whenever possible
An alteration, by definition, changes the design of at least
one component of the vessel shell, and the validity of
the design changes cannot be verified by comprehensive
inspection
An alteration differs significantly from a repair, which does
not involve design changes
The pressure for the hydrotest should be the minimum test
pressure required by the ASME code for name plate design pressure and temperature
The test pressure, after alteration, will normally be lower
than the recommended hydrotest pressure for new vessels, as it is likely that some of the original corrosion allowance will have been consumed during service before
the alteration is made
Hydrotesting pressure vessels that have been altered by the
installation of a new nozzle requiring reinforcement is

sometimes accomplished by welding a cap to the inside of the vessel shell covering the nozzle

This circumvents preparing the entire vessel for hydrotest by providing for a "local hydrotest"

However, a local hydrotest will not develop the same stresses in nozzle reinforcement and the vessel shell component surrounding the opening as would be developed by hydrotesting the entire vessel

The cap will effectively change the shape of the vessel shell component to which it is welded and will have a significant effect on the stresses developed in that component by internal pressure

Thus, a local hydrotest is **not** a valid verification of the design of an alteration, and this practice is **not recommended**

### Approval of Alterations, Documentation, and Name Plate

NB (Paragraph R-502)

Requires that the organization performing the alteration prepare an R-1 form, which must be submitted to an authorized inspector for approval

The organization performing the repair must then attach a new name plate that displays the design pressure (or MAWP) and temperature for the altered vessel

Approval of the alteration and attachment of the new name plate must be obtained from an authorized inspector before the vessel is returned to service

API 510 (Paragraph 7.8.2 (c))

Requires that the documentation of alterations to pressure vessels must be kept as permanent records, but it does not prescribe using a standard form (Paragraph Appendix D)

Approval of an alteration by an authorized inspector is required before the vessel is returned to service, but

attachment of a new name plate is not mandatory unless the design pressure (or MAWP) and temperature
are changed by the alteration

## ► RERATING

### General Considerations

Rerating a pressure vessel consists of changing the design
pressure (or MAWP) and/or temperature from that displayed on the vessel's name plate

Rerating usually **does not** involve a physical alteration of
the pressure-containing capability of the vessel, but can
be required by alterations that are not designed for the
original design pressure and/or temperature

Rerating is **necessitated** most **commonly** by

A change in operating conditions for the process

Deteriorating (i.e., the occurrence of corrosion or cracking) that affects vessel integrity and reliability for the
original design pressure and temperature, and a repair
cannot be economically justified

### Organization Making Rerating

NB (Paragraph R-503)

Requires the rerating of a pressure vessel to be performed
by the original manufacturer whenever possible

Rerating can be performed by a registered professional engineer if the rerating cannot be obtained from the
manufacturer

API 510 (Paragraph 8.2.1)

Permits either the original manufacturer or an experienced engineer employed by the owner/operator to perform the rerating

Only engineers with appropriate experience with pressure vessel design, fabrication, and inspection should perform reratings

A consultant retained by the owner–operator is also acceptable

### Calculations

Rerating a pressure vessel requires making calculations for every major pressure-containing component (i.e., shell, heads, nozzles, reinforcements, and flanges) to **verify** that they **will be adequate** for the new design pressure and temperature

The effect of all internal and external loads on the vessel shell must be considered in the calculations for rerating

Therefore, rerating **involves repeating all the calculations** that were made for the original design of the vessel for the new design pressure and temperature

It can be thought of as designing a pressure vessel in reverse

Instead of calculating the minimum required thickness of each shell component for the prescribed design pressure and temperature, calculations are made to determine if the actual thickness of each shell component is adequate for the rerated pressure and temperature

Both the NB (Paragraph R-503) and the API 510 (Paragraph 8.2.1) require making calculations according to the ASME Code

Decrease in pressure

Rerating of a pressure vessel for a lower pressure is usually required if

The operating temperature is increased for new process conditions

Corrosion has reduced the remaining wall thickness below the minimum required thickness for the original design conditions

Increase in pressure

Rerating of a pressure vessel for a higher pressure can usually be accomplished only if

The operating temperature is decreased

Thickness measurements of all pressure-containing shell components indicate that the original corrosion allowance was greater than necessary for the actual corrosion experienced

Increase or decrease in temperature

An increase in the temperature will almost always require decreasing the pressure, unless the new temperature remains below 450°F

A decrease in temperature will almost always permit an increase in pressure, unless the original temperature was 450°F or below

Rerating for a lower temperature should never be allowed to violate the rules in the ASME Code for low temperature operation

Essential to check the vessel being rerated for compliance with the current rules of the code for low temperature operation when the new temperature will be 120°F or below

This may be very difficult to do when the vessel is old and the materials used for construction are now obsolete

Under these circumstances, it may be necessary to cut samples from the vessel for $C_v$ impact testing to perform a satisfactory rerating

## Information Required

A thorough inspection should be made to assure that the vessel is in satisfactory condition for the new pressure and temperature

It is especially important to determine the minimum remaining thickness of every pressure-containing component of the vessel shell and to detect any cracks that may have developed during service

This will usually require more NDE than normally performed during a routine in-service inspection

## Approval of Rerating, Documentation, and Name Plate

NB treats rerating as an alteration with respect to the requirements for preparation of an R-1 form, approval by an authorized inspector, and attachment of a new name plate displaying the new pressure and/or temperature

API 510 also requires approval of the rerating by an authorized inspector and attachment of a new name plate

The new name plate should be considered mandatory because the pressure and/or temperature for the rerated vessel differs from those displayed on the original name plate

# Index

---

Note: Page numbers followed by *f* indicate figures and *t* indicate tables.

Printed in the United States
By Bookmasters